Goats

Goats

Small-Scale Herding

BY SUE WEAVER

Project Team
Jarelle S. Stein, Editor
Kendra Strey, Assistant Editor
Jill Dupont, Production
Lisa Barfield, Book Design Concept
Michael Vincent Capozzi, Book Design and Layout
Indexed by Rachel Rice

i-5 PUBLISHING, LLC™
Chief Executive Officer: Mark Harris
Chief Financial Officer: Nicole Fabian
Vice President, Chief Content Officer: June Kikuchi
General Manager, i5 Press: Christopher Reggio
Editorial Director, i5 Press: Andrew DePrisco
Art Director, i5 Press: Mary Ann Kahn
Digital General Manager: Melissa Kauffman
Production Director: Laurie Panaggio
Production Manager: Jessica Jaensch
Marketing Director: Lisa MacDonald

The Library of Congress has cataloged an earlier printing as follows:
Weaver, Sue
Goats: small-scale herding for pleasure and profit / by Sue Weaver
p. cm.—(Hobby farms)
Includes index.
ISBN-13: 978-1-931993-67-8
ISBN-10: 1-931993-67-X
1. Goats. 2. Goats—United States. I. Title. II. Series.
SF383.W34 2006
636.39—dc22

2005032283

i-5 Publishing, LLC™
3 Burroughs, Irvine, CA 92618
www.facebook.com/i5press
www.i5publishing.com

Printed and bound in China
17 16 15 14 10 11

This work is dedicated to Karen Keb Acevedo,
my sister in goats, and to Simone, Charlotte,
and Albert, Pygmy goats extraordinaire.

Contents

INTRODUCTION

Why Goats?

Goats were humanity's first domesticated livestock; we've had ten thousand years to get things right. Today's goats provide tasty milk, delicious meat, attractive pelts, and two kinds of renewable fiber. They clear pasture for other livestock by grazing and destroying weeds and brush, they pull carts (goats are amazingly strong), and they pack along the tents and grub when folks go camping. It costs little to buy and maintain goats, and only a modest land plot is required to raise them. Goats are naturals for today's hobby farms.

The world's goat population leapt from 281 million in 1950 to 768 million in 2003; more than 2.5 million of those goats dwell in the United States. The most lucrative livestock venture of the new millennium is raising meat goats—demand by far exceeds supply, and it will for decades to come. Other profitable hobby farm goat ventures include marketing goat's milk and value-added dairy products; mohair and the hides of Angora goats; cashmere; and meat, fiber, and dairy goat breeding stock.

Curious, intelligent, agile, and friendly, goats provide hours of entertainment for their keepers. Everyone who has goats loves them. Whether you want to turn a profit with goats or keep a few for fun, we're here to show you how to get started.

CHAPTER ONE

Goats: A Primer

How long have goats been around? Where did the first ones come from? Are there many different kinds? What are they like? Who raises goats? Before getting into choosing, purchasing, housing, breeding, and other essential subjects, here's a brief look at goats through history and a glance at types, breeds, and traits.

From the Beginning

Goats were domesticated around 8000 BC by the people of Ganj Dareh, a Neolithic village nestled in the Kermanshah Valley of the Zagros Mountains in the highlands of western Iran. According to archaeologists, goat meat had graced the human menu for more than forty thousand years prior to this. The earlier bones gathered from area caves, however, were discards from mature bucks (male goats), the favorites of hunters who needed to bag something big enough to feed a crowd. Toe bones recovered from Ganj Dareh middens are the remains of young bucks, the ones not needed for breeding purposes, and some aged does, females too old to have kids. The change tells us that people had begun keeping goats, rather than just hunting them.

After a one hundred to two hundred–year occupation, the good people of Ganj Dareh packed up their families and possessions, including their goats, and traveled south into the arid Irani lowlands. They resettled away from the wild goat's natural range at a place called Ali Kosh. With a movable food supply—goats and two newly domesticated cereal grains, wheat and barley—humans could abandon their long-time roles as hunter-gatherers and take up the mantle of nomadic herders and tillers of the soil. Archaeological excavations at Jericho unearthed mounds of domestic goat bones carbon-dated to 7000–6000 BC.

Early domestic goats served their human masters exceedingly well. They provided a portable and readily accessible milk and meat supply, fiber for tent covers and clothing, skins for leather, hair-on pelts for robes and rugs, and kids to sacrifice to the gods. Goats packed belongings on their backs and drew travois-type sledges. They were friendly and small, thus easily handled, and required minimal care. Best in arid, semitropical, and mountainous countries, goats survived on browse from trees, brush, and scrub, under conditions in which horses, sheep, or cattle would starve.

Goats spread east from the Fertile Crescent across continental Europe and thence to Great Britain. As elsewhere, goats there became "the poor man's cow," thriving in mountain and moorland crofters' fields and folds, from which they sometimes escaped. Their feral descendants still thrive in remote and isolated pockets along the west coast of Ireland, on Snowdonia in Wales, on Lundy Island and the Isle of Rum, in the Mull of Kintyre, Galloway, and Loch Lomond in Scotland.

During the 1500s, goats came to the Americas with Spanish conquistadors, settlers, and sailors. The Spaniards, like other seafarers of the day, carried aboard their sailing ships this tasty, animated meat supply. It was their custom to salt uninhabited islands with breeding stock, allowing them to harvest future meals on subsequent trips. Historians believe the Pilgrims carried goats on the *Mayflower's* 1620 maiden journey to the New World. Plymouth Colony certainly had them by 1627, when a resident

Domestic goats were a ready source of milk and meat for early settlers.

praised the settlement's goats because "they yeeld commodities with their Flesh, their Milk, their Cheese, the Skinnes, and the Hayre." The Pilgrims considered goat's milk a restorative medicine as well. In the coming centuries, goats accompanied settlers as they pushed westward across North America. By browsing as the party traveled, goats furnished their own eats while providing meat and milk on demand.

By the mid-nineteenth century, generic Spanish goats (also called scrub, brush, hill, briar, and woods goats) could be found in most southeastern states and throughout the Southwest and California. The year 1849 saw the arrival of North America's first purebred goats: seven Angora does and two bucks imported to South Carolina. (Fleece-bearing goats were commonplace in parts of Asia Minor as early as 600 BC.) One of North America's few purely native breeds first made an appearance in the 1880s. An itinerant stranger named John Tinsley came to Marshall County, Tennessee, accompanied by four slightly peculiar goats. When they were startled, their muscles would seize, causing the animals to freeze and sometimes fall over. From these four goats, many believe, emerged the Myotonic goat, a heavy rump breed—with a tendency to topple—popular for meat production and ease of handling.

The 1904 World's Fair in St. Louis, Missouri, heralded a further turning point in goat history when it sponsored the first North American dairy goat

A team of goats stands ready to transport supplies across Alaska in earlier days. As a source of strength and fiber as well as food, the goat played an important role in the settlement of the United States and Canada.

This horned, cou blanc–colored (French, "white neck") French Alpine doe is typical of her breed.

show. The *Missouri Historical Review* noted, "This first provision made at a World's Fair for a display of milch goats brought to the Exposition some choice and home bred specimens." At the same World's Fair, Hagenbeck's Wild Animal Paradise imported two striking Schwartzwald Alpine does and displayed them in a lavish diorama depicting the Alps. This same year the United States formed its first goat registry, the American Milk Goat Record, now the American Dairy Goat Association (ADGA).

In 1906, Mrs. Edward Roby crossed Swiss dairy goats with common stock to develop the American Goat. With them, she strove to supply tuberculosis-free milk to the children of Chicago at a time when many cows were infected. Although she was moderately successful, parents who had never tasted goats' milk refused to give it to their children. During the early 1900s, the first Anglo-Nubians (now simply called Nubians) were shipped from Britain to North America. Between 1893 and 1941, 190 Toggenburg dairy goats were imported; between 1904 and 1922, 160 Saanen. During 1922, the first documented purebred French Alpines, twenty-one in a single importation, arrived by ship, followed in 1936 by five Oberhasli (then called Swiss Alpines). The first documented modern Pygmy goats arrived in North America during the 1950s, originally as novelties to be displayed in zoos. In 1993, the first purebred Boer meat goats, developed in South Africa in the early 1900s, set foot (or hoof) on American soil. Boers took America by storm, as did Kiko meat goats developed in New Zealand and imported at about the same time.

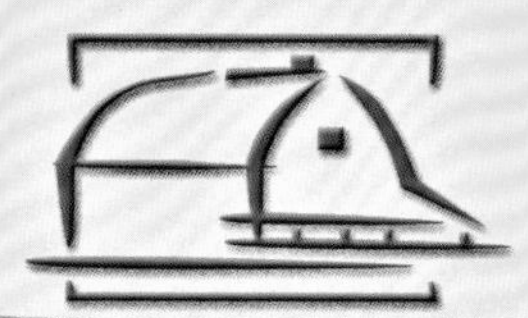

Classic Goats

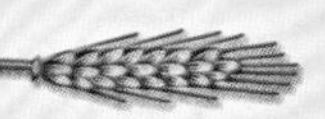

In Words and Images

Goats are mentioned many hundreds of times in sacred books such as the Bible, the Torah, the Koran, and the Bhagavad Gita, often in the guise of sacrifices and as tribute. Goats are pictured on the walls of the oldest known Egyptian tombs and on coins of many ancient realms. A child's toy goat is one of the finest artifacts excavated at India's Harappa ruins, dating to 3000–1500 BC. In 1184 BC, Homer described wonderful goat and sheep milk cheeses—among them forerunners of today's feta—aged in mountain caves in what is now Greece.

Goats were so important to ancient man that some of his deities, spirits, and fairies were assigned goatlike features. Gods Dionysus, Pan, and Silenus had horns and hooves. Fairies and other spirits with goaty features include Greek satyr, Italian faun, Russian ljeschie, Polish polevik, Basque lamiñak, Welsh gwyllion, and the Scottish glastig and urisk. Hindu deities Agni and Kali both rode goats, as did Aphrodite (Greece), Befana (Italy), and Joulupukki (Finland).

Named goats figure in Norse eddas and mythology. The great doe Heidrun gives mead, not milk, for the gods and heroes in Valhalla. Thor's chariot team, Tanngrisnir and Tanngnostr, who pulled his chariot across the sky, could be slaughtered for supper at day's end, then restored overnight. They were ready to head out again at daybreak—a neat trick even for hardy goats!

Goats at a Glance

Domestic goats belong to the Bovidae family, along with other hollow-horned, cloven-hoofed ruminants such as cattle, thence to the Caprinae subfamily, in the company of their cousins, the sheep. Goats are further classified by their genus, Capra, and fall into one of six species: Capra hircus (today's domestic goat), Capra aegagrus (the wild Bezoar goat, ancestor of Capra hircus), Capra ibex (the wild ibex), Capra falconeri (the markhor of central Asia), Capra pyrenaica (the wild Spanish goat of the Pyrenees), and Capra cylindricornis (the Dagestan tur of the Caucasus mountains). (Some scientists divide goats into as many as ten species.)

Roughly one hundred breeds and documented varieties of domestic goats exist in the world today, but fewer than two dozen are available in North America. The world's estimated 768 million goats have many traits in common, including social structure, flocking instincts, and breeding traits.

Goat Classifications

For the goat keeper, goats fall into three basic categories—dairy goats, meat goats, and goats raised for fiber. Subcategories and crossovers certainly exist. Goats are sometimes used to pull carts and pack supplies recreationally and to clear land.

Dairy Goats

Dairy goats are lithe, elegant creatures developed for giving lots of luscious milk. However, excess kids (bucklings not needed for breeding) are often marketed as cabrito (the meat of young kids). Some dairies routinely breed their does to Boer and Kiko bucks to produce

Nubians (called Anglo-Nubians in their native Britain) give less milk than the Swiss dairy breeds produce, but their milk is higher in butterfat. Nubians come in a wide range of colors, and this girl's a knockout with her spotted pattern!

a meatier product. Recreational goat aficionados claim dairy goat wethers, particularly Saanens and Alpines, make the best harness and pack goats bar none.

Dairy breeds readily available throughout North America include the Swiss breeds (Saanens, Sables, Oberhaslis, Toggenburgs, and Alpines), the LaMancha (a distinctly American breed), the Nubian (known in its British homeland as the Anglo-Nubian), and the pint-size Nigerian Dwarf from West Africa. Scaled-down miniature versions of all but Nigerian Dwarfs are out there, too.

An uncommon midsize combination dairy and meat breed, the Kinder goat, was developed by crossing full-size Nubian does with meaty Pygmy bucks. Although Pygmy goats are primarily raised for pets, the does give a surprising volume of high butterfat-content milk.

Meat Goats

Primary purebred meat goat breeds are the immensely popular Boer from South Africa; all-American Myotonics (also known as fainting goats) and their selectively improved counterparts, Tennessee Meat Goats; and the New Zealand Kiko goat. Several exciting combination breeds such as the TexMaster (Boer/Tennessee Meat Goat) and GeneMaster (Boer/Kiko) are being developed, while generic Spanish meat goats form the nucleus of many commercial herds. All are bred for muscle mass, hardiness, adaptability, and exceptional feed-to-flesh conversion ratio. Pygmies are meat goats, too.

Fiber Goats

The backbone of North America's fiber goat industry is the traditional white mohair–producing Angora goat, but the

These muscular MAC Goat full-blood does are shining examples of their breed. Boer goats revolutionized the meat goat industry.

fleece of scarcer-colored Angoras is in high demand for hand spinning, too. A more diminutive fiber producer is the midsize Pygora goat, developed by crossing Angoras and Pygmies. Cashmere goats are the Rolls-Royces of the fiber goat industry, and while uncommon, the American cashmere goat population is growing rapidly.

Recreational Goats

Goats have frequently been driven in harness, sometimes as serious work animals but frequently for recreation. Shortly after their father's presidential inauguration in 1861, Willie and Tad Lincoln were presented with cart goats named Nanny and Nanko. On one occasion, Tad harnessed Nanko to a rocking chair and drove at breakneck speed through a White House reception, causing many a dignified gent and hoop-skirted lady to leap to safety. Most recreational goat buffs prefer wethers, but does, too, can work in harness or under packing gear. A bonus: a lactating pack doe provides fresh, whole milk on the trail. Recreational goat equipment—pack saddles and panniers, carts and driving harnesses—is readily available for goats of all sizes.

Brush Goats

Because goats willingly browse weeds and saplings other animals won't touch—and nourish themselves in the bargain—many people keep them for clearing land of scrub and brush. Dairy does can do the job, but because of potential damage to their large udders, goat keepers prefer not to use them for this particular task. Improved meat goats do well but will usually require supplementary feed. The hands-down champions are hardy generic Spanish goats. They aren't called brush, scrub, and briar goats for nothing.

Eamon, wearing a custom-crafted leather harness, stands ready for cart-pulling duty.

This goat herd moves out under the leadership of the herd queen.

Social Structure

Goats maintained under herd conditions are protected by a single large, strong alpha buck whose role is to breed his choice of does, to maintain discipline, and to guard the group from predators. When the herd moves, members proceed, usually single file. In times of perceived danger the alpha buck protects the rear.

Though other intact males may be part of the herd's hierarchy, these underlings are not permitted to sire offspring. Younger bucks periodically challenge the alpha buck's position. When he's usurped, removed from the herd, or killed through predation, the group readily accepts a new alpha buck.

Not so the herd's true leader, a wise old alpha female, the herd queen. When she moves, all, including the alpha buck, follow. When she halts to browse, everyone eats. Once established, she is herd queen until she's too infirm to do her job

Separating the Sheep from the Goats

- Unless she's ill or frightened, a goat carries her tail up and flipped forward over her back; sheep's tails invariably hang down.
- A goat's horns sweep back from the skull, then upward and sometimes out; most sheep's horns curl back, then down and around into spirals.
- Goats travel widely, gleaning 60 percent of their daily fare from trees and bushes, 40 percent from grass and herbs. Sheep tend to stay closer to home, ingesting 90 percent of daily fare through grazing grass and herbs.
- Goats are a lying-out species: does hide their newborns in the grass or brush as shelter from predators, returning five or six times a day to feed them. Newborn lambs shadow their dams within hours of birth.

Advice from the Farm

Welcome to the Goat World

The experts offer some words of wisdom to new goat keepers.

A Very Friendly Place

"Goats are great, and you will find the goat world is a very friendly place. Figure out what you're looking for in a goat, then find a breed (or mix) that best matches what you want. If you are looking for milkers, choosing older animals is wise as they have been milked a few years and are most likely pretty used to it. They will stand better for you, and you don't have to train them. Have someone give you a milking lesson.

"If you are looking for any old goats, see if there is a rescue near you. Farm animal rescues can be hard to find, but they do get very nice goats that the old owners just couldn't keep anymore. They also take in abused animals, so talk to the rescue to see which would be the best match for you.

"You can get meat goats as kids, you don't need much training with them, but if you're new to raising your own meat, you may get very attached and end up with pets.

"I have Nubians and Myotonic (fainting) goats and I love them both, though I think the fainters are my favorite. Myotonics are meat goats, but I have the small 'pet' size. I have never eaten one.

"Congrats on the new goats you will be getting! Remember that once you start, you will always make room for 'just one more.' "

—Michelle Wilfong

A Lot of Work!

"Dairy goats are a lot of work when they're lactating because they have to be milked twice a day if they're not raising their own kids, and you can't just milk a goat when you feel like it. It must be done on a regular schedule. I've had Nubian dairy goats, and while they're my favorite breed, I just don't have the time to milk.

"Angora and cashmere goats require shearing—twice a year in the case of the Angora—and then what do you with the fleeces? If you're a hand spinner or if you want to market mohair or cashmere on a commercial basis, that's good, but otherwise it's a lot of work! You also need to be careful of the type of pasture you have for fiber goats because it's very easy for them to pick up grass seeds and burrs that will downgrade the quality of their fiber.

"I have pasture-run meat goats of no specific breed. They're relatively easy to take care of. These goats would make good pets if you don't like the idea of selling your goats to the butcher or eating them yourself."

—Glenda Plog

or dies. Confusion reigns until members select a new herd queen—often one of the former queen's daughters.

A typical day goes something like this: Come daybreak, the herd rises and sets out to browse, its two stalwart rulers in the lead. Herd queen spies a tasty stand of blackberry brambles. She stops to nibble leaves. The rest of the herd crowds around and feeds, too. Herd king eats but remains alert. When herd queen feels it's time to move along, she gives herd king a long, hard look. When he notices, he heads out with his queen, and the herd follows. When it's time to stop and digest, queen gives king another look. He leads them to a sheltered area where herd queen picks a cushy spot, and they all lie down and ruminate.

Older goats boss younger ones, big goats lord it over smaller peers. Horns account for a lot. A female goat with horns generally ranks higher than her hornless sisters, and alpha bucks with impressive horns are challenged less .

Biological Traits*

Rectal temperature: 101.5–104.5 degrees Fahrenheit
Pulse: 60–80 beats per minute*
Respiration: 12–25 breaths per minute*
Ruminal Movements: 1–1.5 per minute
Blood Capacity: half to third of body weight
Natural Life Span: 10–12 years (well-maintained goats have lived 20 years or more)
Sight: Goats have relatively poor depth perception but otherwise keen sight. They distinguish certain colors and unless fleece obstructs their vision, they take in a 280- to 320-degree visual field.
Taste: Goats distinguish between bitter, sweet, salty, and sour tastes. Their high tolerance for bitter flavors allows them to savor vegetation species that other ruminants won't eat.
Hearing: Goats have very acute hearing, encompassing a full sound spectrum from high to very low pitch.
Teeth: Mature goats sport four pairs of incisors on the lower jaw (a hard structure call a dental palate or dental pad takes the place of upper incisors), plus three premolars and three molars on each side of the upper and lower jaws.

* parameters run slightly higher for kids

Breeding Traits

Depending on breed and condition, bucks reach sexual maturity at three to ten months. Does require six months to a year. Because occasional precocious kids mature faster than the norm, goat keepers separate the sexes by twelve to fourteen weeks. One mature buck can impregnate as many as fifty does in a sixty-day breeding season.

Most goats breed seasonally from early fall through late winter, though breeds developed in hot climates, such as Boers and Kikos can cycle (come into heat or estrus) and conceive year-round. Goats cycle every eighteen to twenty-two days and remain receptive to the buck for twelve to thirty-six hours; ovulation generally occurs during the last hours of standing heat. Depending on age and breed, gestation ranges from 148 to 156 days and leads to the birth of one to four (or more) kids.

CHAPTER TWO

A Buyer's Guide to Goats

Don't rush out to buy some goats. It's a bad idea when purchasing any type of livestock but especially risky when getting into goats. Though goats aren't hothouse flowers, neither are they the happy-go-lucky, can-noshing species of movies and cartoons. Goats require specialized handling and feeding—and keeping goats contained in fences is never a lark. Goats are cute, personable, charming, and imminently entertaining. They can be profitable, particularly in a hobby farm setting. But goats are also destructive (picture a four-legged, cloven-hoofed, tap dancer auditioning on the hood of your truck), mischievous, sometimes ornery, and often exasperating. Be certain you know what you're getting into before you commit.

Find yourself a mentor. Most experienced goat producers are happy to teach new owners the ropes. To track down a mentor, ask your county extension agent for the names of owners in your locale, join a state or regional goat club, or subscribe to goat-oriented magazines and e-mail groups to find goat-savvy folks in your area. A mentor or extension agent can talk with you about which breed will meet your needs and what to look for when buying your goats and what happens once you do. You need to educate yourself as well. Here are the issues you should consider and the basic information you should have on goat-buying transactions.

Choosing the Breeds

Before going goat shopping, know precisely what you want. Make a list of the qualities you're looking for, star the ones you feel are essential, and note which ones you're willing to forgo. Some breeds fare better than others in certain climates. Certain breeds are flighty. Some make dandy cart goats, whereas others are too small for harness work unless you plan to drive a team. If you want a goat who milks a gallon a day, a Pygmy doe won't do. However, if you're looking for a nice caprine friend and you don't want to make cheese or yogurt, a Pygmy doe (or two) could prove the perfect choice. (See box "Common Goat Breeds in Brief.") Consider availability as well in your choice—whether you're willing to go farther afield to get exactly the breed you want.

Purebred, Experimental, Grade, or American?

Registered goats generally cost more to buy than do grade (unregistered) goats, but you might not need to spring for registered stock. It depends on your goals. If you plan to exhibit your animals at high-profile shows, or to sell breeding stock to other people, you probably do. If you want a pack wether, a 4-H show goat, or a nice doe to provide household dairy products, registration papers aren't essential.

A registration certificate is an official document proving that the animal in question is duly recorded in the herdbook of an appropriate registry association. Depending on which registry issues the certificate, the document will provide a host of pertinent details, including the goat's registered name and identification specifics—such as its birth date, its breeder, its current and former owners, and its pedigree. Dairy breed papers also document milk production records in great detail. You can contact the ADGA with any questions you may have about the latter.

Wee baby Salem, just three weeks old, is three-fourths Boer and one-fourth Nubian, a popular type of percentage Boer goat. His famous sire is the MAC Goats champion buck Hoss.

The four categories of dairy goats in terms of registration are purebred, experimental, grade, and Americans. *Purebreds* are registered goats that come from registered parents of the same breed and have no unknowns in their pedigrees. *Experimentals* are registered goats that come from registered parents but of two different breeds. A goat of unknown ancestry is considered a *grade*. However, several generations of breeding grade does to ADGA-registered bucks (always of the same breed) and listing the offspring with ADGA as *recorded grades* eventually results in fully registerable *American* offspring. For example, seven-eighths Alpine and one-eighth grade doe is an American Alpine; a fifteen-sixteenths Nubian and one-sixteenth grade buck is an American Nubian. However, ADGA terminology doesn't apply to meat goats.

This is Morgan, our sweet Sable baby bred by Christie's Caprines. Saanens have occasionally produced colored offspring, called Sables, which recently have come to be recognized as a separate breed.

To qualify as a registered *full-blood* in the American Boer Goat Association herdbook, all of a goat's ancestors must be *full-blood* Boer goats. Registered *percentage* does are 50 to 88 percent full-blood Boer genetics; *percentage* bucks are 50 to 95 percent Boer. Beyond that (94 percent for does, 97 percent for bucks), they become *purebred* Boers. *Purebreds* never achieve *full-blood* status.

The International Kiko Goat Association registers *New Zealand full-bloods* (from 100 percent imported New Zealand bloodlines), *American premier full-bloods* (of 99.44 percent or greater New Zealand genetics), *purebreds* (87.5 to 99.44 percent New Zealand genetics), and *percentages* (50 and 75 percent New Zealand Kiko genetics). To avoid making costly mistakes, learn your breed's registration lingo before you buy!

Pets, cart and pack goats, brush clearers, and low-production household dairy goats needn't be of any specific breed. Mixed-blood goats cost less to buy and no more to maintain than fancy registered stock and may be precisely the animals you need.

Availability

If you're seeking Nubians, Pygmies, or Boers, you'll probably find a plentiful supply of good ones close to home. Less common breeds, such as Sables, Kinder goats, and colored Angoras, may be a

Goat auctions and buying stations such as this one are marketing mainstays for commercial meat goat producers.

different story. If you don't want to travel long distances to buy foundation or replacement stock, pick a common breed or at least one popular in your locale. Conversely, though it takes more effort to start with something out of the ordinary, it also assures a market for your goats—other seekers don't want to range afar, either.

Purchasing goats from a distance has its pitfalls because you may not be able to visit the sellers and inspect potential purchases in person. If this is the case, buy only from breeders whose sterling reputations (and guarantees) take some of the gamble out of long-distance transactions. The transportation of distant purchases is also an issue, but it needn't be a major one. Livestock haulers and some horse transporters carry goats cross-country for a fee. Kids and smaller goats can be inexpensively and safely shipped by air.

If you're buying close to home, you can locate breeders via classified ads (free-distribution classifieds are especially rich picking), through notices on bulletin boards (watch for them at the vet's office and feed stores), and by word of mouth (your county extension agent or vet can usually put you in touch with local goat owners). Or place "want to buy" ads and notices of your own.

To get a feel for breeders and to learn what sort of goats they have for sale, visit breed association Web sites or subscribe to print and online goat periodicals. Peruse the ads and breeders directories, and sign up for goat-oriented e-mail groups.

Goats auctioned through upscale production sales and consignment sales hosted by bona fide goat organizations are generally the cream of the caprine crop. Never buy goats at generic livestock sale barns. Run-of-the-mill livestock auctions are the goat farmer's dumping ground. Most animals run through these sales are culls or sick, and the ones who aren't will be stressed and exposed to disease. A single livestock sale bargain can bring nasties the likes of foot rot, sore mouth, and caseous lymphadenitis (CL) home to roost, sometimes to the tune of thousands of dollars in vet bills and losses. Buy your goats through high-profile goat auctions or from private individuals.

Selecting the Goats

The cardinal rule when buying goats: start with good ones. Choose the best and the healthiest foundation stock you can afford.

Conformation

Acceptable conformation—defined as the way an animal is put together—varies among dairy, meat, and fiber goats. It's important to study a copy of your breed's standard of excellence, available from whichever registry issues its registration papers, before you buy. Don't discount the importance of good conformation; you'll pay more for a correct foundation goat, but he's worth it. Even if you never show your goats, buyers will pay higher prices for your stock.

Health

Never knowingly buy a sick goat! Carefully evaluate potential purchases

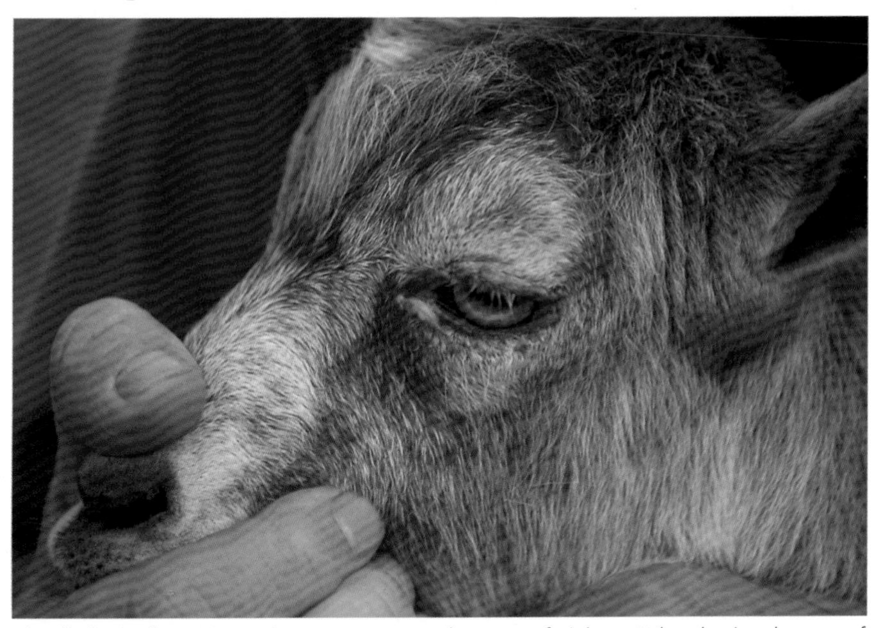

The discharge from Morgan's eyes suggests early stages of pinkeye. When buying, beware of goats with runny eyes; there could be a serious health issue. Fortunately, Morgan's problem was simply dust irritation and was easily treated with saline solution and antibiotic eye ointment.

before bringing them home. A healthy goat is alert. He's sociable; even semi-wild goats show interest in new faces. A goat standing off by himself, head down, disinterested in what's going on is probably sick or soon will be.

A healthy goat is neither tubby nor scrawny. He shows interest in food if it's offered, and when resting, he chews his cud. His skin is soft and supple; his coat is shiny. His eyes are bright and clear. Runny eyes and a snotty nose are red flags, as are wheezing, coughing, and diarrhea (a healthy goat's droppings are dry and firm). Unexplained lumps, stiff joints, swellings, and bare patches in the coat spell trouble. Avoid a limping goat; he could have foot rot (or worse).

If in doubt and you really want a particular animal, ask the seller if you can hire a vet to take a look, and consider it money well spent.

Morgan is a polled Sable, meaning he was born without horn buds. The lumps on his forehead show where his horns would have been.

Horns

If you don't like horned goats, don't buy a goat that has them; you can't simply saw them off. The cores inside a goat's horns are rich in nerves and blood vessels. Dehorning, even done by a veterinarian and under anesthesia, is a grisly, dangerous, and ultimately painful procedure that leaves gaping holes in an animal's skull. With dedicated follow-up care these holes will eventually close, but why expose an animal to this kind of torment?

Dairy goat kids are routinely disbudded when they're a few days to a week or so old. This is accomplished by destroying a kid's emerging horn buds, burning them with a disbudding iron. Though it's painful and not a procedure best performed by beginning goat keepers, disbudding is far more humane than exposing a goat to full-scale dehorning later on.

Meat and fiber goat producers and recreational goat owners are far less likely to eschew horns, but all goats exhibited in 4-H shows—even the ones that are shown in 4-H meat goat, fiber goat, driving, and packing classes—must be hornless or shown with blunted horns.

Should horns be a problem? It depends. You probably don't want them if you confine your goats (they'll butt one another, probably causing injuries); if they'll be expected to use stanchions or milking stands; if you have small children who might get poked, or if you'd prefer not to be poked yourself; if your other goats are polled (naturally hornless) or disbudded. However, science theorizes that horns act as thermal cooling devices, so if you have working pack or harness goats or you live where it's hot, they're a boon.

Teeth

A goat has front teeth only in the lower jaw. In lieu of upper incisors, there is a tough, hard pad of tissue called a dental palate. For maximum browsing efficiency, the lower incisors must align with the leading edge of the dental palate, neither protruding beyond it (a condition called monkey mouth or sow mouth) nor meeting appreciatively behind the dental palate's forward edge (parrot mouth).

Beginning at about age five, a goat's permanent teeth begin to spread wider apart at the gum line, then break off, and eventually fall out. A goat with missing teeth is said to be broken-mouthed. When his last tooth is shed (around age ten), he's a gummer. Aged goats with broken teeth have difficulty browsing, so unless you're willing to feed soft hay or concentrates, check those teeth before you buy.

Sex-Specific Factors

No matter what class of stock you raise—be they dairy, meat, or fiber goats—buy does with good udders. A goat's udder should be soft, wide, and round, with good attachments front and rear. The two sides should be symmetrical. Avoid lopsided, pendulous udders with enormous sausage teats, especially in dairy goats, and reject goats with extremely hot, hard, or lumpy udders—these being telltale signs of mastitis involvement.

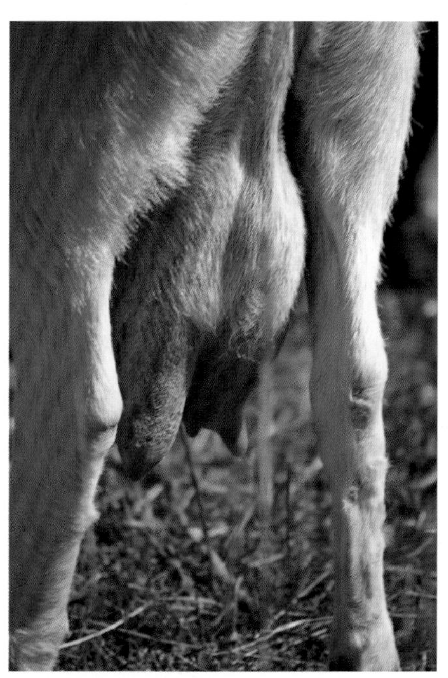

Note the enlarged left teat of this goat. Lopsided udders are undesirable.

Dairy goats should have two functioning teats with one orifice apiece. Deviations from the norm are serious faults and are rare. Dairy kids are sometimes born with additional vestigial teats, but they're usually removed when doelings are disbudded.

Meat goats, especially Boers, are often graced with more than two teats. In Boers, up to two adequately spaced, functional teats per side are acceptable. However, nubs (small, knoblike lumps

This Oberhasli's scrotum is just right. When buying a buck, size counts; large testicles equate with fertility and breeding vigor.

that lack orifices), fishtail teats (two teats with a single stem), antler teats (a single teat with several branches), clusters (several small teats bunched together), and kalbas or gourd teats (larger roundish lumps that have orifices) frequently occur. A blind teat (one lacking an orifice) can be dangerous if newborns consistently suckle on it in lieu of a functional one; the kids will literally starve. Most of these irregularities disqualify a doe from showing.

Male goats have tiny teats, too; they're situated just in front of the scrotum on a buck. Although they aren't important in and of themselves, check for the same irregularities in breeding bucks as you would in does. Bucks with unacceptable teat structure may sire daughters with bad udders. Bucks with more than two separated teats per side generally can't be shown.

Bucks must have two large, symmetrical testicles. When palpated, the testicles should feel smooth, resilient, and free of lumps. An excessive split separating the testicles at the apex of the scrotum (more than an inch in most breeds) is unacceptable. When choosing a buck, size matters. The greater his scrotal circumference, the higher his libido and the more semen he'll likely produce. A mature buck of most full-size breeds should tape 10 inches or more, measured around the widest part of his scrotum. Boer bucks must tape at least 11.5 inches (American Boer Goat Association) or 12 inches (International Boer Goat Association) by maturity at two years of age.

When buying a wether, ask when the goat was castrated. Since castration abruptly halts the development of a young male's urinary tract and affects adult penis size, early castration predisposes male goats to water belly, also known as urinary calculi. In this condition, mineral crystals in his urine block his underdeveloped urethra and cause his bladder to burst; death occurs within a few days. Castration of pet and recreational goats is best postponed until the animal is at least one month old (later is better).

Whichever sex you're considering, be aware of one of the peculiarities of goat breeding: breeding polled goats to one another sometimes results in her-

Matt Gurn shows a field of friendly MAC Goats Boers to visiting buyers. Goats are curious; these crowd around to see what's going on.

maphrodite offspring (displaying both male and female sexual organs). It pays to check, keeping in mind that male goats always have teats, so you don't end up with one these unusual goats.

The Sale

You've done your homework, and you're ready to buy. Based on your research, contact sellers who produce the sort of goats you want. Make appointments to visit and view their animals. Goat producers and goat dairy farmers are busy folks, so keep your appointments or call to cancel. It never hurts to ask for a seller's references in advance, especially when buying expensive goats. Be sure to check them out before your visit.

When you arrive, look around. Though fancy facilities are never a must, goats should be kept in clean, safe, comfortable surroundings. Do the goats appear healthy? Are they friendly? Are their hooves neatly trimmed? Their drinking water clean, their feeders free of droppings? Evaluate the seller, too. Does he or she seem knowledgeable, honest, and sincere?

Ask to see prospective purchases' health, worming, and breeding production records (and milk production records for dairy goats). Virtually all responsible goat breeders and dairy operators keep meticulous records. If the seller can't produce them, be suspicious.

Carefully inspect paperwork when buying registered goats. Have registration certificates been transferred into the seller's name? (He can't legally transfer them into your ownership unless he's the certified owner of record.) Does the description on the papers match the goat? Check ID numbers tattooed inside ears (and sometimes the underside of their tails) against numbers printed on registration papers, ditto numbers embossed on any ear tags. Sometimes a seller has "misplaced the papers" and will "mail them to you when they turn

Common Goat Breeds in Brief

Here's a brief look at the different breeds of goats you can choose from depending on whether you want dairy, fiber, or meat goats, or pets.

Dairy Goats

Alpine (also called French Alpine)

Alpine goats originated in the French Alps. They are medium to large goats—does at least 30 inches tall and 135 pounds, and bucks 34 inches and 170 pounds. Friendly, inquisitive Alpines come in a range of colors and shadings. Because of their productivity and good natures, Alpines are popular in commercial dairy settings.

LaMancha

The almost-earless LaMancha (at least 28 inches and 130 pounds) is an all-American goat developed in Oregon during the 1930s. Goat fanciers claim LaManchas are the friendliest of the dairy goat breeds. They can be any color. Two types of ears occur among them: gopher (1 inch or less in length, with little or no cartilage) and elf (2 inches or less in length, with cartilage). LaManchas produce copious amounts of high-butterfat milk.

Miniature Dairy Goats

The Miniature Dairy Goat Association registers scaled-down (20–25 inches tall, weight varies by breed) versions of all standard dairy goat breeds, among them Mini-Alpines, Mini-LaManchas (MiniManchas), Mini-Nubians, Mini-Oberhaslis, Mini-Saanens, and Mini-Toggenburgs. Miniatures have the same standards of perfection as those of full-size counterpart breeds.

Nigerian Dwarf

Nigerian Dwarfs are perfectly proportioned miniature dairy goats, capable of milking three to four pounds of 6–10 percent butterfat per day. Gentle, personable Nigerians can be any color. They breed year-round; multiple births are common. (Four per litter is the average; though there have been births of as many as seven.) Does are typically 17–19 inches tall, bucks 19–20 inches; 75 pounds for both sexes.

Nubian (also called Anglo-Nubian)

Nubians were developed in nineteenth-century England by crossing British does with bucks of African and Indian origins. A noisy, active, medium- to large-size dairy goat (does at least 30 inches and 135 pounds, bucks 35 inches and 175 pounds), Nubians are known for their high-butterfat milk production, sturdy build, long floppy ears, and aristocratic Roman-nosed faces. All colors and patterns are equally valued.

Oberhasli

Alert and active, Swiss Oberhaslis are medium-size goats (minimum for does is 28 inches and 120 pounds, for bucks

30 inches and 150 pounds). They are always light to reddish brown accented with two black stripes down the face, a black muzzle, a black dorsal stripe from forehead to tail, a black belly and udder, and black legs below the knees and hocks.

Saanen

These big (30–35 inches and 130–170 pounds) solid white, pink-skinned dairy goats from Switzerland are friendly, outgoing heavy milkers, with long lactations. They are popular commercial dairy goats, often called "the Holsteins of the goat world."

Sable

Sables are colored Saanens, newly recognized as a separate breed. Because their skin is pigmented, they don't sun burn as Saanens sometimes do.

Toggenburg

Toggs are smaller than the other Swiss dairy breeds. They are some shade of brown with white markings (white ears with a dark spot in middle of each, two white stripes down the face, hind legs white from hocks to hooves, forelegs white from knees down).

Fiber Goats

Angora

The quintessential fiber goats, Angoras produce long, silky, white or colored mohair. Angoras are medium-size goats (does are 70–110 pounds, bucks 180–225; height varies). They aren't as hardy as most other breeds. Twinning is relatively uncommon. Angoras must be shorn at least once a year.

Cashmere

Cashmere goats are a type, not a breed. Goats of all breeds, except Angoras (and one class of Pygoras), produce cashmere undercoats in varied quantities and qualities. High-quality, volume producers are considered cashmere goats.

Pygora

Pygoras were developed by crossing registered Angora and Pygmy goats. They're small (does at least 18 inches tall and 65–75 pounds; bucks and wethers at least 23 inches tall and 75–95 pounds), easygoing, and friendly, and they come in many colors. Some

Pygoras produce mohair, some cashmere, and others a combination.

Meat Goats

Boer

The word *boer* means "farmer" in South Africa, land of the Boer goat's birth. Big (does weigh 200–225 pounds and bucks 240–300 pounds; height can vary greatly), flop-eared, Roman-nosed, and wrinkled, the Boer is America's favorite meat goat. Boers are prolific, normally producing two to four kids per kidding, and they breed out of season, making three kiddings in two years possible. Boer colors include traditional (white with red head), black traditional (white with black head), paint (spotted), red, and black.

GeneMaster

GeneMaster goats are three-eighths Kiko and five-eighths Boer goats developed by New Zealand's Goatex Group company, the folks who pioneered the Kiko goat. Pedigree International currently maintains the North American GeneMaster herdbook.

Kalahari Red

Kalahari Reds look like large, dark red Boers. Kalahari Reds are a developing breed in South Africa. Though a few American producers are breeding true South African stock, most North American "Kalahari Reds" are simply solid red Boers.

Kiko

Kiko means "meat" in Maori. Kikos were developed in New Zealand by the Goatex Group. Beginning with feral goat stock, breeders selected for meatiness, survivability, parasite resistance, and foraging ability and, in doing so, created today's ultrahardy Kiko goat.

Myotonic

Today's Myotonic goats (also called fainting goats, fainters, wooden legs, Tennessee Peg Legs, and nervous goats) are believed to be the descendants of a group of Myotonic goats brought to Tennessee around 1880. When these goats are frightened, a genetic fluke causes their muscles to temporarily seize up; if they're off balance when this happens, they fall down. Myotonic goats come in all sizes and colors (black and white is especially common). They don't jump well, so they're easy to contain; and they're noted for their sunny dispositions.

Savanna

Big, white, and wrinkled, South African Savanna goats resemble their Boer cousins, but with a twist. South African Savanna breeders used indigenous white goat foundation stock and natural selection to create a hardier-than-Boers breed of heat-tolerant, drought-and-parasite-resistant, extremely fertile meat goats with short, all-white hair and black skin. Savannas' thick, pliable skin yields an important

secondary cash crop: their pelts are favorites in the leather trade. Fewer than a score of North American breeders offer full-blood Savanna breeding stock, but interest in the breed is skyrocketing. Pedigree International maintains the official Savanna herdbook.

Spanish

Spanish goat is a catchall term for brush goats of unknown ancestry, so no breed standard exists. Spanish goats can be any color, although solid white is most common; both sexes have huge, outspreading horns.

Tennessee Meat Goat

Suzanne W. Gasparotto of Onion Creek Ranch developed the spectacular Tennessee Meat Goat by selectively breeding full-blood Myotonic goats for muscle mass and size. Pedigree International maintains the Tennessee Meat Goat registry.

TexMaster Meat Goat

The TexMaster Meat Goat, another Onion Creek Ranch development, was originally engineered by crossing Myotonic and Tennessee Meat Goat bucks with full-blood and percentage Boer does (meaning they are a only certain percent Boer, not 100 percent). Pedigree International keeps its herdbook as well.

Other Breeds

Kinder

The Kinder goat (does 20–26 inches, bucks 28 inches; weight varies) is a dual-purpose milk and meat breed developed by crossing Nubian does with Pygmy goat bucks. Prolific (most does produce three to five kids per litter) and easygoing, Kinders make ideal hobby farm milk goats and pets.

Pygmy

Nowadays, Pygmy goats (does are 16–22 inches, bucks 16–23; weight varies) are usually kept as pets, but they developed in West Africa as dual-purpose meat and milk goats. Pygmies are short, squat, and sweet natured. Lactating does give up to two quarts of rich, high-butterfat milk per day, making Pygmies respectable small-family milk goats.

ALBC Conservation Priority List Breeds

The American Livestock Breed Conservancy (ALBC) includes six goat breeds on its Conservation Priority List. Two require immediate help: the critically endangered San Clemente of relatively pure Spanish stock, and the threatened Tennessee Fainting (also called the Myotonic goat or fainting goat). Listed also: Spanish (under the Watch category), Nigerian Dwarf and Oberhasli (Recovering), and another distinctly American product, the scarce, island-bred Arapaw goat (Study). (See the Resources section for ALBC contact information.)

Advice from the Farm

Let the Buyer Beware

Our experts share tips about goat buying.

Hit the Books

"Get some books on goat health. These are great references. They will scare you because they'll list everything that can go wrong. But until you've been raising goats for thirty or forty years, you won't see even half of those things and even then, you'll still be learning things about goats."

—Rikke D. Giles

Beware the Bargains

"Before you get a goat, read all you can about goats and talk to people who have them. Start with an older doe or wether and *then* get a kid. And don't buy goats at sale barns. Animals are usually sold at auction for a reason. Sometimes you can get a decent animal if you know what you're looking for, you know the people who consigned the animal, and you're lucky. The people that own the sale barn near me do not like goats, so they're all put into the same pen: bucks, does (very pregnant or dry), little kids, big goats, and little goats—then they're exposed to all sorts of diseases (pinkeye, snotty noses, abscesses, and so on). If you do buy a goat at a sale barn, don't put it with your others until it's been in quarantine for at least two weeks."

—Pat Smith

Don't Take Any Lumps

"Although I raise five breeds of pet and show goats and two of sheep, I look for many of the same qualities I'd look for if I were buying meat or dairy goats. I want healthy goats. I look for clear eyes, moist noses, shining coats, strong straight backs with level toplines. I also look for strong, straight legs that don't have spun hocks or knobby knees. Seeing an animal run helps assure me it has healthy legs. I look at goats' berries to see if their color and formation show good internal health. I'm also looking for lumps, abscess, crooked jaws, herniated navels, or cleft palates."

—Bobbie Milsom

Go for the Goat!

"I have a milk cow and milk goats. You only need two goats for them to be happy and content, and you can keep seven head of goats per one cow. Plus, goats are more intelligent, friendlier, and safer."

—Samantha Kennedy

Did You Know?

According to United Nations statistics, the world's goat population grew from 281 million in 1950 to 768 million in 2003. The world's top ten goat-producing countries are China (172.9 million), India (124.5 million), Pakistan (52.8 million), Sudan (40 million), Bangladesh (34.5 million), Nigeria (27 million), Iran (26 million), Indonesia (13.2 million), Tanzania (12.5 million), and Mali (11.4 million).

up." Don't buy the story. Without an up-to-date registration certificate in your hand, you're paying registered price for a goat that may be grade.

Judgments based on intuition aren't always accurate, but if you feel uncomfortable with any part of a seller's presentation, seek elsewhere.

After the Sale

Sellers will often deliver your goats for a modest fee; it's the easiest way to get your purchases home. You can, of course, fetch them yourself if you prefer or if the seller doesn't deliver. Diminutive goats such as kids and adults of some miniature breeds are easily transported in high-impact plastic airline-style dog crates stowed in the bed of a truck (secured directly behind the cab to block wind), in a van, or in an SUV. Horse trailers, stock trailers, pipe racks, and topper-clad truck beds all suffice. Whatever you use, bed the conveyance deeply for the animals' comfort, and use tarps to keep goats out of direct wind and drafts.

Goats mustn't be stressed in transit; stress equates with serious, sometimes fatal, digestive upsets. Keep everything low-key. Avoid crowding. Provide hay to nibble en route, stop frequently to offer clean drinking water, and dose your goats with a rumen-friendly probiotic paste or gel such as Probios or Fast Track before departure and after you reach your destination.

Have facilities ready to receive your goats, and feed them the same sort of hay and concentrates to which they're accustomed. Many sellers will provide a few days' feed for departing goats if you ask. Begin mixing the old feed with the new feed to help the goats gradually make the change. You won't want to further stress newcomers by immediately switching feeds.

Isolate newcomers from established goats or sheep (goats and sheep share many diseases) for at least three weeks. Deworm them on arrival, and if their vaccination history is uncertain, revaccinate as soon as you can.

Kari Trampas's LaMancha buck is a sterling example of his breed, famous for sunny dispositions and high butterfat milk. They originated in California, making them America's own dairy goat breed.

CHAPTER THREE

Housing and Feeding Your Goats

Build your goats a showplace barn, and they'll love it. Or hammer together a three-sided shanty built of recycled lumber and secondhand corrugated roofing—and they'll love it. Given a cozy, dry place to sleep in a draft-free shelter, goats are content. They're the essence of simplicity to house. Feeding is easy, too, once you've learned the basic rules. It's important, though, not to make common mistakes, and we'll show you how to avoid them.

THE RIGHT HOUSING

Goats hate being wet. Trees and hedges can provide sufficient shade from light showers, but goats in rainy and snowy climates need access to weather-resistant, man-made structures, too. In most climates, a three-sided structure (sometimes called a loafing shed or a field shelter), with its open side facing away from prevailing winds, makes an ideal, inexpensive goat shelter. Other basic shelters include movable A-frames crafted of plywood; commercial calf hutches; hoop structures designed for hogs; straw buildings; and even large, prefabricated doghouses. Bucks are hard on housing; they bash, butt, climb and scratch their surroundings. Build buck shelters, pens, and fences out of stout, sturdy materials.

If you breed goats, you'll need enclosed housing. Close-to-term does, does with newborn kids, and delicate bottle kids require dry, draft-free housing, especially during the harsh winter months. Dairy goat owners also need covered, weather-resistant areas in which to set up their milking stands. If need (or preference) dictates keeping your goats in confinement housing, you'll probably want to house them in a barn.

Bucks, such as this young Savanna intent on demolishing his fence, can be terribly destructive. Construct bull-proof buck enclosures!

Goats like to see one another. Consider making interior pens out of pipe or heavy-duty welded wire panels in lieu of solid walls. Goats also love to climb. Elevated sleeping platforms make for happy campers, as do playgrounds built of recycled telephone cable spools, slanted walk-upon climbing planks, and elevated perches. Provide getaways where kids or low-ranking herd members can escape aggressors; airline-style dog crates are effective hideouts for your little ones, and pens with narrow openings will provide refuge from big, bad bullies.

No matter what type of housing you construct, you have to consider the basics necessary to ensure health and comfort: space, drainage, ventilation, flooring, and bedding. You also need to ensure that your herd has the right feeding and drinking accommodations. Last but not least, you must determine the best means for containing and protecting your charges.

The Structural Basics

Whether fashioning quarters to house a single 4-H goat or a vast herd of meat goats, allow at least 15 square feet of bedded floor space per goat. Make certain drainage is adequate, and slope the roof away from the shelter's open side so rain and snow cascade off the rear, rather than the front, of the structure.

When building field shelters for small numbers of animals, keep the roof height as low as you can. Five to 6 feet in front, 3 to 4 feet in back is just about right. Low-slung roofs hold body heat at dozing-goat level, essential in colder northern climates. The disadvantage: squat buildings are harder to clean, especially if you do the chore by hand.

No matter a structure's size, goat housing must be adequately ventilated.

Saanen and Sable kids at Christie's Caprines find shelter in a shade house made of cast-off pallets. During the heat of summer, goats need shade.

Goats housed in damp, poorly vented barns are prone to serious respiratory ailments, as are goats (especially kids) exposed to drafts.

In most climates, packed dirt or clay floors are better than cold, hard concrete. In arid climates, wooden floors work well, too, but eventually they'll rot, necessitating replacement. Whatever the flooring, bed the structure with 4 to 6 inches of absorbent material, such as straw, discarded hay, wood shavings or sawdust, peanut hulls, ground corncobs, or sand. Although some goat owners clean and rebed indoor stalls daily, many prefer deep-litter bedding. With deep-litter bedding, you continually add just enough bedding to keep floors dry, only periodically (a few times a year is usually adequate) cleaning everything out, down to floor level. This system is both comfortable and warm and is extremely simple to maintain. Whatever material you use, you'll need to find a responsible way to dispose of it when it becomes soiled. Compost it, give it away, sell it, but especially on nonagricultural, zoned properties, don't let it pile up.

If you keep your goats confined, they'll need a safely fenced, communal exercise area, one allowing at least 30 square feet of space per goat. Mixing horned and hornless goats in a pasture setting can work, but not in close quarters where considerably more jostling and sparring occurs. Whenever possible, horned and disbudded or polled goats should be housed and penned separately.

Troughs and Feeders

Goats require copious supplies of fresh, clean water, kept reasonably cool in the summertime and liquid when tempera-

Salem sips out of a reused cattle lick tub. Used mineral lick tubs make primo water troughs.

tures dip below freezing. Install running water and electricity in your barn or shelter, or locate the structure within easy garden hose and extension cord reach of existing utilities. If you do have electrical wiring, it must be protected with conduit and kept well out of curious goats' reach. Any glass windows should be protected by screens.

Provide multiple watering troughs or buckets in lieu of a single big one. It's infinitely easier to dump, scrub, and disinfect several smaller containers than it is a full-size trough. If one water source becomes contaminated with nanny berries, there will be others for your goats to choose from.

Goats fed on the ground are prone to disease and excessive internal parasite infestation, and they just plain waste a lot of feed by trampling on it and soiling it. (Goats won't touch soiled feed.) Buy commercial goat feeders or build your own. Prime requisites are poop-proof hayracks and feeders, which must be installed higher than your tallest goat's tail or at least be easily cleaned, be difficult for kids to climb up into, and be designed so goats can reach their feed but not get their heads stuck while doing so. You'll find lots of easy and effective feeder construction plans free on the Internet. (Since sheep feeders work well for polled and disbudded goats, search for plans for goat hay feeders and sheep hay feeders.)

Don't store feed where goats can help themselves. Overeating, especially of grain or of rich legume hay, can quickly kill the toughest goat. Store grain in goat-proof covered containers with snug lids. (Fifty-five-gallon food-grade plastic or metal drums and decommissioned freezers work well.) Secure the feed room door with a goat-proof lock; opening hook and eye closures are child's play to a nimble-lipped goat. You may need to add a padlock or something similar.

A neighbor's doe is caught in a woven wire fence. Horned goats often put their heads places they shouldn't. Had no one come along to free her, this doe might have starved or been killed by a predator.

Fences

The cardinal rule of goat keeping, especially if you plan to pasture your animals: don't buy goats until you've erected stout, goat-resistant fences. Goats are curious, mischievous creatures. When the urge to wander strikes them, your goats will do their level best to escape. They love going walkabout in the countryside (especially the roads), and they live to clog on your car. Your neighbor's prize tea roses are as yummy as wild ones, and oh, garden veggies taste good!

Goats can squeeze through incredibly small gaps, so standard plank and post fences won't phase them unless lined with good wire fencing. Woven wire, also called field fence or field mesh, makes terrific perimeter, paddock, and pen fencing and can be used to render wooded fences goat tight.

Woven wire fencing consists of smooth, horizontal wires held apart by vertical wires called stays. It's sold in regular galvanized, high-tensile, and colored polymer-coated high-tensile versions; with verticals placed at 6- to 12-inch spacings (wider verticals prevent horned goats and sheep from getting their horns caught), in 26- to 52-inch heights. Horizontal wire spacing generally increases as the fence gets taller. When buying woven wire, check the numbers: 8/32/9 fencing has 8 horizontal wires, it's 32 inches tall, and it has vertical stays every 9 inches.

Disadvantages of woven wire are its cost and the time and effort required to install it. The advantages are it's safe, it looks good, and it requires very little upkeep once it's properly installed. Most goat producers use 32-inch woven wire and stretch several strands of barbed or electrified high-tensile smooth wire above it for extra height. An offset electric wire installed inside the fence at adult goat shoulder height prevents rubbing by itchy goats and helps keep livestock guardian dogs inside the fence.

Ultra-sturdy, welded wire livestock panels can be used instead of woven wire, albeit at much greater cost. We like it for smaller areas, such as buck runs, paddocks, and pens, where its ease of installation and sturdy, no-maintenance construction offset installation costs. One caveat: most welded wire panels have jagged, sharp, snipped-wire edges.

Heavy-welded wire cattle panels are being used here to goat-proof a galvanized farm gate. Such panels work well as goat fencing.

To prevent tearing your clothes (and your goats' eyes and hides), use a heavy-duty rasp to smooth them.

Barbed wire, the traditional farm fencing, can be effective when built using eight to ten strands of evenly spaced, tightly stretched 15-gauge or better wire, preferably augmented by add-on twisted wire stays installed between posts. Installing barbed wire is not for the faint of heart, and when run into, it can cause catastrophic injuries to man and beast. (It should never be used where horses are pastured or ridden.) Yet it's relatively inexpensive and is often used to upgrade existing farm fences to goat-resistant status. It's marketed in galvanized, high-tensile, and polymer-coated versions in various gauges, distances between barbs, and barb lengths. It's available at all farm stores.

Smooth, electrified wire fencing makes inexpensive, effective goat fencing, although a goat-resistant fence charger will set you back some bucks. Farm stores stock electric wire of numerous types and gauges, but for the long haul, high-tensile versions work best. To make it work for goats, you'll need to install stout braces and end posts and use six to nine strands of wire. Train your goats to respect the electric fence by luring them to it with grain until they get zapped. High-tensile versions should not be used to fence pastures shared with horses.

Other types of electric fencing worth investigating are electrified string-net, polywire, polytape, and rope fences designed to be used with step-in posts. Since they're the essence of simplicity to move, they're ideal for dividing pastures and for temporary grazing settings, and most styles are remarkably goat-proof. One serious drawback: goats, especially horned goats, can become tangled in

Pretty, petite, Alpine baby Atticus nibbles on a tasty blade of hay.

net-style temporary electric fencing, panic, and struggle until they die. Another disadvantage is that if you live in hard, rocky terrain, using step-in posts is probably not an option.

Goats Don't Eat Tin Cans

Feeding goats is a complex subject best discussed with your county extension agent or a livestock nutritionist who can suggest forage, concentrates, and supplements based on your goats' ages and breeds, on whether they're pregnant or lactating, and on the types of feedstuffs available in your locale. However, certain truths apply no matter where you live or what sorts of goats you have to feed.

Ruminate on This

Goats are ruminants, as are sheep, cattle, and deer; their digestive systems are very unlike those of simple-stomached species such as horses, carnivores, and humans. In lieu of a single stomach, every goat has a rumen, reticulum, omasum, and abomasum. Each compartment has a specialized job to perform.

In newborn kids, only the abomasum is functional. When a kid raises her head to nurse, a band of tissue called the esophageal groove closes and shunts milk directly from her esophagus to her abomasum. That's why it's important for bottle kids to be fed at doe-teat height (see chapter 6). As a kid suckles her dam's udder and begins nibbling plants, dirt, and the rest of her environment, she ingests the microbes she needs to

Before Building Anything!

Before building goat structures (or renovating existing facilities) and before installing goat fencing, scope out applicable zoning laws and touch base with your county extension agent. Your agent will understand your needs and can provide housing bulletins and plans specific to your climate and location. You can also visit the housing and fencing pages at University of Maryland Cooperative Extension's Maryland Small Ruminant Web site. Anyone installing wire fences should also peruse the University of Missouri Extension Service's online bulletin, "Selecting Wire Fencing Materials." See the Resources section for Web sites and other helpful sources before buying fence material and supplies.

Our goats love this fine-stemmed Bermuda grass hay. Good forage should be the mainstay of every caprine diet.

This bale's solid yellow color tells a tale: it was probably rained on after it was mown. Not a good choice for goat munchies.

kick-start rumen function. At three to six weeks of age, she's functioning like a grown-up goat.

The rumen, located on the goat's left side, is the largest (and first) of the four stomach compartments. The rumen does not secrete digestive enzymes—it's essentially a fermentation vat housing the vast horde of friendly microorganisms that convert cellulose into digestible proteins.

As author Suzanne W. Gasparotto of Onion Creek Ranch succinctly puts it, "You are not raising goats, you are raising rumens." When rumen microbes stop processing cellulose for their host, she will sicken and sometimes die. So it's important to learn to assess rumen health. Check to see if a goat is chewing her cud. Use a stethoscope to listen at the goat's bulging left side. A healthy rumen "rumbles" every forty-five to sixty seconds, depending on time of day and what the animal has eaten. Listen to goats in every situation—goats who are healthy and those who are sick, goats who have recently eaten and those who have not. Note how much their left sides bulge. Then you'll know when trouble is brewing and be in a position to help.

When a healthy goat eats, she quickly tanks up on whatever looks tasty, then retires to "chew her cud." She burps up semimacerated material from her rumen, rechews it, then swallows it again. She continues the process until her ruminal microbes have digested the food enough for it to pass into and through the reticulum (where certain nutrients are absorbed) and on to the omasum. The omasum decreases the size of food particles and removes

Poisonous Plants

Goats, (Don't) Pick Your Poison

Though goats safely process most plants, there are some that they simply ought not to ingest. These plants range from being mildly toxic, meaning they must be eaten over a period of time to cause problems, to killing with a single mouthful. These are some to watch out for.

- Amaryllis *
- Apricot (wilted leaves)
- Avocado (leaves, fruit)
- Azalea *
- Baneberry *
- Bindweed *
- Bitterweed *
- Black Henbane *
- Black Locust (bark, seeds, new growth)
- Black Snakeroot *
- Black Walnut *
- Bleeding Heart *
- Bloodroot *
- Bracken Fern *
- Buckeye *
- Buttercup *
- Caladium *
- Calla Lily *
- Cherry (all varieties, wilted leaves)
- Clematis *
- Crocus *
- Crow Poison *
- Daffodil *
- Daphne *
- Death Camas *
- Dieffenbachia *
- Dogbane *
- Dutchman's-Breeches *
- Elephant Ear *
- English Ivy *
- English Laurel *
- Flax *
- Foxglove *
- Horse Nettle *
- Horsetail *
- Hyacinth *
- Hydrangea *
- Iris *
- Jimsonweed *
- Johnsongrass *
- Jonquil *
- Lantana *
- Larkspur *
- Lily-of-the-Valley *
- Lobelia *
- Locoweed *
- Lupine *
- Mayapple *
- Milkweed (leaves)
- Mistletoe *
- Monkshood *
- Mountain Laurel *
- Narcissus *
- Nightshade *
- Oleander *
- Peach (wilted leaves)
- Plum (wilted leaves)
- Poison Hemlock *
- Pokeweed (seeds)
- Potato (leaves, stems)
- Privet (berries)
- Rattlebox *
- Red Maple (leaves)
- Rhododendron *
- Rhubarb (leaves)
- Scotch Broom *
- Sneezeweed *
- Tomato (leaves, stems)
- Water Hemlock *
- White Snakeroot *
- Wisteria (seeds and pods)
- Yellow Jessamine *
- Yew *

* All parts of these plants are toxic or poisonous

Symptoms of plant poisoning include dilated pupils, teeth grinding, vomiting, labored breathing, cries of pain, racing or weak pulse, bloating, scours, muscular weakness or tremors, staggering gait, hyperexcitability, and convulsions.

If you suspect plant poisoning, remove your goat's feed and make her comfortable, supply her with lots of clean drinking water, and get her to a vet as soon as possible, taking along samples of any suspected poisons.

Ask your county extension agent which poisonous plants grow in your locale or visit the "Poisonous Plants and Other Plant Toxins" page at the University of Maryland Cooperative Extension's Maryland Small Ruminant Web site (http://sheepandgoat.com/poison.html) for links to bulletins covering your state or region.

excess fluid from the mix. Finally, the material moves to the abomasum, where body enzymes effect final digestion.

Let Them Eat Forage

The millions of microbes (bacteria, protozoa, and other microorganisms) that populate your goats' rumens and digest whatever they eat require mainly cellulose fiber, meaning forage (browse, hay, or grass), to survive. Concentrates such as commercial goat feeds, corn, and other grains ferment more rapidly than does forage, producing excess acid that can readily kill both the beneficial microbes and your goat. The bulk of all goats' diets must be forage, supplemented by concentrates only when the goats truly need them.

The best dry forage is long-fiber grass hay. High-protein hays, such as alfalfa, clover, and lespedeza, cause the same serious problems as high-protein concentrates: urinary calculi, acidosis, bloat, founder, milk fever, and ketosis (in pregnant does). Goats tolerate (and even savor) a weedier mix than many species do, but all hay must be green, sweet smelling, and absolutely dust and mold free.

This easy-to-build, homemade mineral feeder is constructed of PVC pipe and fittings. After seeing this one at MAC Goats, we built some for our goats and sheep.

A salt lick pan keeps this salt block out of the dirt. Goats need daily access to salt and minerals.

Dairy does, late gestation and nursing does of all breeds, and most fast-growing young stock require grain. Choose clean, mold-free, commercial goat or horse mixes and cracked or whole cereal grains, and store them (and hay) where birds, cats, and wildlife won't contaminate the feed with disease-carrying droppings.

Did You Know?

The forage and feed that a dairy doe eats and inhales can flavor the milk she gives you. According to the University of California Cooperative Extension's bulletin "Milk Quality and Flavor," 80 percent of the off-flavors in goat's milk are feed related.

The best way to avoid objectionable flavors is to eliminate moldy hay and grain, grub suspected plants from your pastures, and remove certain feedstuffs at least five hours prior to milking.

Plants known to flavor milk include bitterweed, buckthorn, buttercups, wild carrot, chamomile, cocklebur, cress, daisies, fennel, flax, wild garlic and onions, horseradish, wild lettuce, marigold, mustards, pepperwort, ragweed, sneezeweed, and yarrow. Feeds best put out at least five hours before milking include alfalfa, soybeans, rye, rape, turnips, cabbage and kale, and clover.

Toxins from certain poisonous plants, when ingested by cattle or goats, end up in your milk supply and can be a threat to you and your family. Abraham Lincoln's mother died after drinking white snakeroot–tainted cow's milk. It can happen today, so be careful.

To supplement the diet, provide a high-quality mineral mix or lick formulated for your type of goats and your locale. Place licks and mixes where goats won't inadvertently poop on them. Goat products generally include copper in quantities that are toxic to sheep, so if you keep both, it's vitally important that you choose a dual-species (low copper) mix or place goat minerals where your sheep can't reach them.

Goats are extremely selective eaters. Given the option, they'll nibble choice bits of hay and dump less-savory morsels onto the floor or ground, where they'll eventually poop and pee on them. Most goats would starve before eating soiled hay, so plan on feeding from waste-resistant feeders and using discarded hay for bedding or feeding goat discards to less-picky species, such as cattle and horses.

Dos and Don'ts

Dietary changes must be made over a period of time—no exceptions. Abrupt changes trigger serious digestive upsets that will kill your goats. You must establish a routine, and stick to it. Don't stress your goats by skipping or delaying their feedings.

Allow enough hayrack and feeder space for all goats in a group to eat at the same time (12–16 inches of feeder space per goat is usually sufficient), and keep feeders clean. Goats won't (and, for health reasons, shouldn't) eat or drink from fouled hayracks, feeders, and water sources. Make certain each goat is eat-

This herd of commercial meat goats waters from a typical Ozark stock pond. Goats require access to plentiful supplies of cool, clean water.

ing. Goats who refuse their usual feed are probably ill.

Do not allow your goats to get fat. It's not healthy. Goats don't marble fat throughout their bodies the way most species do; it's deposited around their internal organs, where, in large quantities, it can inhibit vital function. Learn to assess body condition, keeping in mind that there will be individuals who are leaner or chunkier than the norm.

Pure, Clear Water

The cheapest, most essential nutrient of all is water. Goats won't thrive without 24-7 access to lots of sparkling clean, good-tasting water. They need it to maintain digestive health. Lactating does require water to make milk, and without it males form urinary calculi. Don't skimp. Keep those tanks and buckets filled and clean. Consider installing an automatic watering fixture. Your goats are sure to love you if you do.

Did You Know?

When humans hear the word *pasture*, they visualize meadows of lush, waving grasses. Not so our friend the goat. Like their cousins the deer, goats prefer twigs and leaves and wildflowers and weeds—with perhaps a spot of grass to round out the menu. They thrive on brush-studded rough and rocky land.

Furthermore, since they prefer different plants, goats can be pastured with livestock such as horses and cattle. Goats not only avoid the grasses these other animals require but also grub out invasive weeds, briars, saplings, and brush, clearing land for grassland pasture.

Property owners appreciate goats' taste for such hard-to-rout nasties as purple knapweed, wild blackberry, leafy spurge, purple loosestrife, musk thistle, and multiflora rose. Some entrepreneurs rent goats to landowners specifically to grub out brush and weeds.

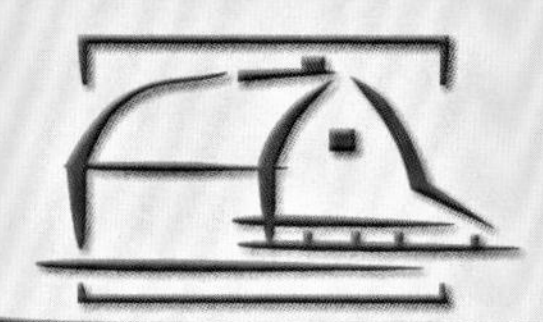

Advice from the Farm

Keeping Them Happy, Healthy, and Warm

Our experts share housing and fencing observations.

Goats Don't Care About Fancy

"I have one large and two small goat structures, and my goats never use the small ones since they like to stick together. In the summer, they sometimes lounge in the shade of the buildings if it gets over 100 degrees and they are full. But most of the time, no matter how hot it gets, they lie along the ditch bank since it provides them with a higher place from which to survey the world.

"I chuckle when I see people put up big, fancy goat houses as they always end up telling me, 'Well gosh, it was a waste of money, I should have just put up a lean-to.'

"A friend of mine just moved to Texas and bought a nice little ranch with an air-conditioned barn. She thought her goats would be in heaven, but they are terrified of the slight noise it makes and won't go into the barn. She has to shut off the air conditioner. Then they'll go in, but they don't spend much time because it's enclosed, and without air-conditioning it gets so hot.

"Goats hate change. I decided to put a fan in my milking parlor to cool it down. Boy, that was a huge mistake. My does wouldn't come into the parlor for four days. I had to chase my nice, gentle milk goats down and drag them into it. They got used to it, but now I have to make sure it is going before they step foot into the barn or they freak out because they can't hear it running. Goats!

"Just remember the number one thing: goats hate getting wet. Make sure the building you put up is waterproof."

—Samantha Kennedy

Simple—But Convenient

"Housing can be a very basic structure depending on where you live. A three-sided shed would be fine if you don't get bad winters. In colder climates, it's nice to be able to shut the door and windows in really bad weather. As long as the goats have enough room to get in and not be crowded, it doesn't need to be huge.

"Don't store your feed where they can get into it. A separate building (or, if you have a larger barn, a separate room) is best.

"It's nice to be able to feed goats without having to go in with them. If you have a half wall, or a walkway they can stick their heads into, you can feed without being run over."

—Michelle Wilfong

CHAPTER FOUR

Goat Behavior and You

Goats are intelligent creatures. Exactly how intelligent is uncertain. Goats (like cats) spurn IQ tests devised by humans and test much lower than they should. In fact, a goat usually prefers to do things his own way. Consider the interesting similarity between the words *caprine* (relating to goats) and *caprice*. *The American Heritage Dictionary of the English Language* defines *capricious* as "characterized by or subject to whim; impulsive and unpredictable." That, in just a few words, describes goats.

The same goat on different days and under different circumstances can seem as smart as a whiz kid or as dense as a box of brick. He may lope down the driveway behind your truck screaming in anguish because you're leaving him or crouch silently under the feed shed when he senses you want to give him a shot.

Goats are at all times nimble and curious, a combination that means trouble unless you (and your neighbors) have goat-proof fences and finely tuned senses of humor. Goats do get out. Goats fandango on truck hoods, clamor up slanted roofs, and unlatch complex locks securing gates. Sashaying down the highway at 2 a.m. is quite the goatish thing, as is raiding your neighbor's lush garden.

Follow the Leaders

Instinct plays a major role in the lives and loves of goats. Ten thousand years after they were domesticated, they still possess intrinsic knowledge of the ways of the wild goat herd.

As discussed in chapter 1, like most species, goats maintain a pecking order. Top goat eats or sleeps whenever and wherever he or she pleases. Second-ranking

This doe spits out a plant she has just sampled. The herd queen utilizes this technique for showing the rest of the gang what not to eat—though she makes a bigger production of it for emphasis.

goat defers to top goat, but lords it over the rest of his herd mates. And so it goes, down to the goat at the bottom of the heap, who eats last, gets the worst place to sleep, and has to jump when any other goat says boo. In the wild, each flock is led by a herd king and a herd queen. On the farm, you will probably have only the latter.

The herd queen, a wise old doe, will lead until she dies or becomes too infirm to carry out her duties. She's respected by all; underlings rarely jostle for her position. At her passing, confusion reigns until a new herd queen (frequently a daughter of the old queen) takes her place.

The herd queen shows the others what and what not to eat. No one samples a plant before the queen. If she eats, they eat the same thing. If she tastes something nasty, she makes a grand production of spitting, sputtering, and wiping her mouth on the earth.

A certain amount of jostling for position in the herd inevitably takes place, especially among newcomers and bachelor males. Fighting is a one-on-one proposition; goats don't gang up on a single opponent, although they may figuratively stand in line for their chances to trounce a newcomer. Aggression occurs between both sexes, including wethers (castrated males), and consists of staring; fluffing of the coat (as the hackles rise), particularly along the spine; front foot stomping; pushing; rushing; horn threats (chin down with horns jutting forward); and outright warfare. Goats don't back up, then charge head-down as rams do. An aggressive, annoyed goat positions himself or herself at a right angle to the opponent's body with head facing the adversary. When ready, the

This broken chamois–colored Alpine doe ("broken" meaning her coat is splashed with white) races across the field at Pat Smith's Anvil Acres. Goats run fast, a fact verifiable by anyone who has ever attempted to catch a runaway goat.

attacker rears onto hind legs and pivots, swooping forward and down, usually smashing into the opponent's flank, neck, or head.

Spend some time with your goats. You can easily tell which doe is your group's herd queen. You must kindly but firmly make it clear that she must defer to you so she'll accept you as her better. As acting herd queen (no matter your sex), you can get your goats to follow wherever you lead them.

Get a Handle

When moving goats or handling them for procedures such as worming and inoculations, it's important to understand how your goats are likely to react and why. Being your goats' herd queen can be essential here because it's all but impossible to drive a herd of goats. Goats lack the keen flocking instinct that causes frightened sheep to mob together and move as a single mass. A startled group of goats scatters, although individuals may whirl to face what spooked them. If still alarmed, they bolt, calling on their speed and agility to outmaneuver perceived danger. Goats are a herding dog's worst nightmare.

Goats dislike entering or crossing water or areas of deep shadow, they resist passing through narrow openings, and only with great reluctance will they attempt to navigate slippery surfaces. They readily move forward out of darkness toward light, from confinement toward open spaces, and into the wind rather than downwind, and they'll more readily go uphill than down.

Loud noises and sudden movements frighten them. Startled or dis-

The collar this Toggenburg cross is wearing allows her owner to easily lead and restrain her without resorting to grabbing those magnificent horns.

gruntled individuals are apt to lie down to avoid being driven or handled. They also aggressively butt and shove at the animals around them.

Most goats resist if you haul them around by their horns. It's better to lead or restrain a tame goat (even a horned one) by his collar or to temporarily immobilize him by cupping a hand under his chin and lifting his head. If you do handle a goat by the horns, avoid snapping off a tip (which will bleed like crazy) by grasping the horns down by the bases.

Harried goats are easily stressed. Since stress invariably leads to serious health problems, avoid stressing them as much as you can. Keep things low key when dealing with goats, and don't lose your temper. Give goats a chance to understand what you're asking before you react; patience goes a long way in the goat yard.

Don't underestimate the power of a goat. When handling seldom-handled goats or when working with goats in close quarters, wear long-sleeved shirts, long pants, and boots or steel-toed shoes. Keep small children out of the action altogether when working around fractious goats.

Goats are smart and have long memories. Depending on the actions you take today, things will be better (or a whole lot worse) the next time you handle your goats.

Can You Read Me Now?

Goats communicate mainly through body language, but sometimes they

This spotted beauty is Kari Trampas's Nubian buck. The bucks at her Seymour, Missouri, farm share a single pen.

vocalize as well. An alarmed goat stands rigid, poised to run but with legs firmly planted, his tail curled over his back, his head held high and ears pricked forward at perceived danger. He may stamp one forefoot or snort to alert the herd. His alarm snort resembles a loud, high-pitched sneeze. To assert their authority, dominant individuals glare at, crowd, bite, and butt underlings. High-ranking, assertive individuals may try this with you. Nip aggression in the bud. You always need to be top goat in your herd.

Although goat vocalization traditionally has been called bleating, the sounds goats make are increasingly being referred to as calls. Goats call in greeting (to their human caretakers and to other goats), to demand food, and to locate their kids and other herd members. Does murmur tenderly to their newborn kids, and goats scream in terror and in pain. Some breeds call more than others do. Nubians are the noisiest of all.

The Birds and the Bees and Behavior Keys

Breeding season brings a new set of behaviors to the goat yard, some of them peculiar by anyone's standards.

It's a Guy Thing

As breeding time approaches (see chapter 6), a buck goes into rut. He hopes to attract the ladies, and what better way than by liberally dousing himself with "perfume"? Unless he's been descented, the glands on his forehead begin exuding a pungent, earthy musk. He adds to his allure by spritzing his face, beard, chest, and belly with thin jets of urine. He also grasps his penis in his mouth

Two rare Savanna does enjoy the rugged terrain of Carl and Shirley Langle's Diamond L Ranch in Viola, Arkansas.

(yes, he is that agile) and sometimes urinates, whereupon he curls his lips in a grimacelike response (a behavior formally called flehmen).

And this before he meets a doe in heat!

When he's turned out with the ladies, he'll add new tricks to his repertoire. He'll trail a prospective girlfriend, sniffing her sides and under her tail, sometimes pawing her with a stick-rigid front leg, all the while flapping his tongue and making bizarre vocalizations called blubbering. When she pees, he samples it and flehmens. When she stops running from him and stands to be bred, he mounts her. As he ejaculates, he flings back his head, then recuperates for a heartbeat before dismounting.

During rut, bucks become more pugnacious, even toward humans. Being rammed by a buck is more than a playful bunt in the butt. A big buck can knock you over and seriously hurt you; so can a little one if he happens to clip you behind both knees.

Bucks penned apart from the ladies practice their techniques among themselves. Some bucks, especially bottle-raised and pet bucks, may court their favorite humans. This isn't much of a problem when a 40-pound Nigerian Dwarf wants to rub his forehead glands on your leg, but it's a serious one when a 300-pound Boer wants to mount your twelve-year-old daughter. It's important to stay alert when working around bucks in rut. Children and vulnerable adults shouldn't handle them at all.

Does Just Want to Have Fun

Does have their own set of unique breeding behaviors, although some overlap with those of bucks. Females

Advice from the Farm

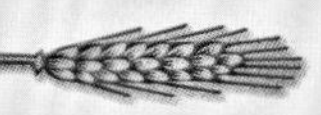

Goat Antics

Our experts give you the inside scoop on goat behavior.

8622. Une vieille Chevrière cévenole

The LaMancha Death Glare

"I have a LaBoer (LaMancha/Boer) goat named Olive, who is extremely aggressive toward women. She doesn't love women; she sees us as competition. She stands on her hind legs and paws at me and has the LaMancha Death Glare going on; she is quite intimidating. I actually have to catch her up when other women come to look at the goats. She really loves my husband and boys, but boy, does she get jealous of me! If they are out with me, she runs the fence bawling until they go over and pet her; then she is fine. If she didn't give me such gorgeous kids, I would eat her!"

—Samantha Kennedy

The Ego Check

"Just about the time you think you understand goats, why they behave in such a manner or what is needed to keep them healthy, one will come along to blow your theories out of the water. Just like in psychology class, 'Is it nature or nurture?' and in nursing school, 'Not everyone reacts the same to a drug.' They have a knack for keeping your ego in check (just like my own kids did). The only thing that I can vouch for is that I continue to learn something new with each goat. When I stop enjoying the learning process it will be time to do something different."

—Mona Enderli

The Storm Trooper Bounce

"You may be able to 'potty train' a goat for the house, but boy when mine get in, which they manage to do sometimes through the pet door, they are wild! They get on the desk, the TV, up in the chairs, just nutso! They come bouncing in like storm troopers! It takes an act of Congress to get them gathered up and back out, and there are only three of them.

"Outdoors, they are all over my car, and they have boards to walk on and concrete blocks to climb on. I had a 7-foot ladder leaning against the house, and yesterday one of them climbed partway up and it fell over and knocked the window out of the front door! They are exasperating! I don't know if I can wait till fall or not to get the fencing done."

—Donna Haas

Eamon stands quietly attached to a hitching rail, licking his mouth in contentment. The use of treats for training is very effective with goats.

don't pee on themselves, but tongue flapping, blubbering, and mounting female herd mates (and their human caretakers) are not uncommon. This is referred to as "being bucky."

Other signs of estrus (heat) include allowing other does to mount them, frequent urination, decreased appetite, clear mucous discharge from the vulva, tail-wagging (flagging), a great deal of strident calling, and mood changes. The super-sweet doe may intentionally kick over the milk pail or attack her underling herd mates, while old Picklepuss wants to love you to death. A lactating doe's milk production takes a dramatic dive while she's in heat, too.

Train Them Up

If you're a large-scale meat producer, you probably don't need to train your goats, except possibly for the bucks. (Because you'll probably keep them in separate quarters part of the year and handle them more often than you handle your does, it's always a good idea to educate bucks.) If you call them while rattling a bucket, they'll stampede to your handling facilities. That may be enough training for your purposes. However, if you keep a few pet, 4-H, or recreational goats, you'll at least want to teach them to lead and tie (remain quietly attached to a hitching rail). Whatever your training goals, there are some points to keep in mind.

Goats work their hearts out for food, making them ideal candidates for clicker training. Clicker training, also known as operant conditioning, is widely used to train sea mammals, horses, and dogs. Horse training methods can be easily tailored for schooling goats. If you've never tried clicker training before, we recom-

To encourage a goat to enter the ring with his young exhibitor, the owner gives him a quick shove on the rump. This is the most effective method to use when a goat falls behind or stops.

mend starting by reading one of the clicker training books listed in the Resources section. Although you probably won't want to teach your goat to fetch a soda from the fridge (then again, you might), most of the training routines common with other pets work exceptionally well with goats.

Reward-based training always works best, but when you need to thwart undesirable behavior immediately, reach for a high-powered water gun or a household pump sprayer with a long, strong jet. Goats despise water, especially when it's squirted in their faces. A loud "No!" coupled with a blast or two of water tends to grab the most errant goat's attention. Don't just yell and wave your arms and chase your goat away. To goats, chasing is play behavior; that means you're actually rewarding the goat for misbehaving.

Goats can be led using a halter or a collar. A halter tends to give you more control. Walk with your goat's shoulder at your right hip; until he understands, ask someone to follow and urge him along when he falls back or stops. A well-timed, brief shove on his rump works better than pushing him or swatting him with a switch. Reward him when he does well. He'll learn much faster if he's having fun.

For safety's sake, never use choke-type collars or slip-style halters to tie up your goat. Use a slipknot in your rope so you can untie him quickly if he pulls back or somehow gets tangled. *Don't* go off and leave a semitrained goat tied up. You need to be Johnny-on-the-spot to save him if he panics.

CHAPTER FIVE

Goat Health, Maladies, and Hooves

Goats are prone to a host of serious ailments, and that's a fact. However, it's just as true that properly managed goats rarely get sick. We've said it before, but we'll say it again: don't buy trouble; choose healthy foundation stock and take basic steps to watch over the health of your charges. Certain problems will still be inevitable. Anyone who raises kids eventually battles coccidiosis. Pneumonia tends to rear its head from time to time. Nutritional maladies such as bloat, pregnancy ketosis, and goat polio aren't uncommon, especially while newbie goat keepers learn to properly feed their caprine charges. However, it's easy to avoid major nasties such as hoofrot, CAE (caprine arthritis encephalitis), CLA (caseous lymphadenitis), CE (contagious ecthyma, also called sore mouth and orf), and Johne's disease by buying from disease-free herds. At the very least, avoid poor risk purchases from livestock auctions or poorly managed herds of thin, limping, abscess-ridden sheep or goats. (Sheep and goats contract many of the same diseases and parasites.)

Once you have purchased your goats, you should have relatively few health concerns to contend with if you do the following: feed and handle your goats properly, worm them when needed, vaccinate according to your vet's recommendations, monitor their health every day, and don't allow them contact with anyone else's goats or sheep.

Anytime you take your goats where other goats or sheep are present, you're exposing them to a host of communicable diseases. If the risk is necessary or unavoidable, at least hedge your bets against introducing sickness into your herd by quarantining incoming goats—be they newly purchased or home from an out-

Most major feed companies market nutritionally balanced, bagged goat concentrates and pelleted dewormers such as these.

ing—for at least three weeks. To prevent accidental contamination, don't feed and handle quarantined goats until after you've seen to your main herd, and don't let visitors, pets, or poultry move freely between your quarantine area and your primary goat housing.

It's important to recognize illnesses and start treatment right away, so you'll have to learn to tell when a goat is feeling out of round. Stroll among your goats at least twice a day, watching for signs of illness, and remove suspect individuals to your quarantine area without delay. If only one goat is ailing, reduce stress (which takes a heavy toll on already compromised individuals) by moving a second goat to the quarantine area and penning her a distance from, but within sight of, the sick goat.

Recognizing Maladies

Although discussing all the maladies that befall goats is beyond the scope of this book, you'll find the most important ones in "Goat Diseases at a Glance" in the Appendix. Before identifying what ails your goat, you need to be able to recognize when there's something wrong. The chart at right will help you know what to look for.

Assembling Your Resources

So you think you have a sick goat. You've isolated the animal, but you're not quite sure what's wrong. Call your vet immediately. Time is of the essence when treating problems such as coccidiosis, pneumonia, pregnancy, and toxemia, as well as a host of other serious caprine complaints.

Sometimes, however, a goat-savvy vet is temporarily out of pocket. You need to have other people and resources to consult about the situation. Start (before trouble arises) by finding a mentor, an experienced, local goat owner

Know the Signs

A Healthy Goat	A Sick Goat
Alert, lively, and curious about surroundings.	Little or no interest in surroundings; stands or lies alone, away from the other goats.
Tail is carried gaily, level with (most dairy breeds) or up and over (Boers and most other meat breeds) the back.	Head and tail droop.
Normal interest in food; actively chews cud.	Nibbles or refuses food, and may not chew cud.
Bright, clear eyes, free of discharge; nose cool and dry; no thick nasal discharge (scant amount of thin, clear discharge usually all right).	Eyes dull, often squinting; thick yellow or greenish discharge oozes from eyes or nose. Grinds teeth (indicating pain).
Regular, unlabored breathing.	Labored breathing and coughing (congestion); rapid, shallow breathing (more than 20–25 breaths per minute in adults). Slobbers; unusually sweet-smelling breath (ketosis).
Calls in usual timbre and tone.	Call tone and timbre unusual; grumbling or crying out in pain.
Wide, well-developed barrel, indicating well-developed rumen. Body neither blubbery fat nor snake thin; "just right" for breed or type.	Slab-sided (poor rumen development); barrel painfully distended (bloat).
Coat clean and glossy; no obvious knots or abscesses anywhere on body.	Coat dull and unthrifty; scabs, abscesses, and bare patches. Continually rubbing, scratching, or biting self. Ruffles coat, with spine hackles standing on end.
Doe's udder reasonably symmetrical; warm and soft, not hard and icy cold or fiery hot.	Unusually hot or cold, swollen, or painful udder producing thick, nasty-smelling, or clotted milk (mastitis).
Most males urinate relatively infrequently; when they do, they urinate effortlessly and in a steady stream.	Male goat cries while frequently attempting to urinate; produces only a dribble or nothing (likely urinary calculi—requires immediate treatment).
Droppings firm and pelleted.	Droppings extremely smelly, in liquid form tinged with blood or mucus; any change in color or consistency is suspect.
Movement fluid; no unusual hitches or limps; no abnormally swollen joints; no rank-smelling discharge from hooves.	Moves poorly or not at all; hunched over, limping; grossly swollen joints; knee-walking in front (CAE or more likely, hoofrot).
Rectal temperature—101.5°F to 105.5°F.	Rectal temperature lower or higher than norm.

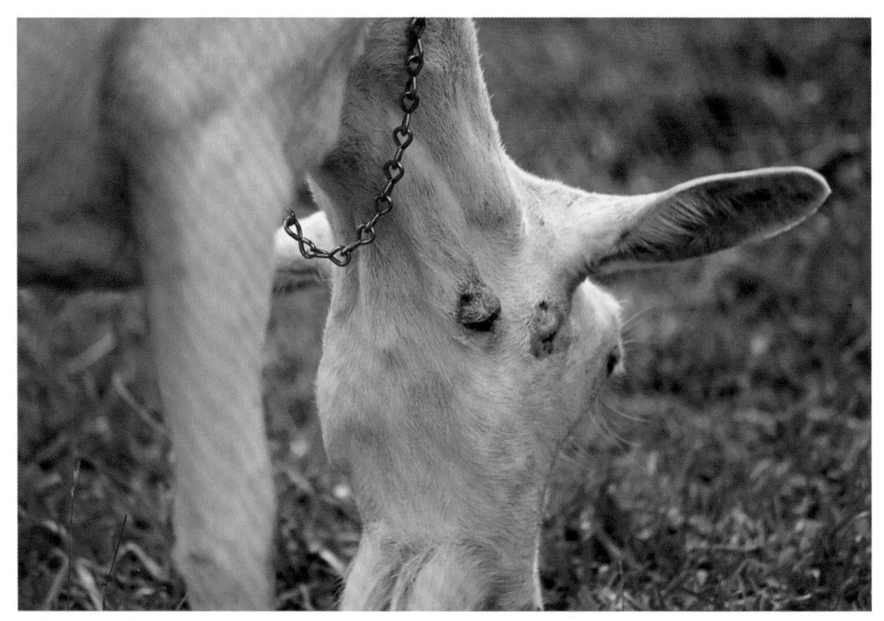

Is it caseous lymphadenitis? It's best to have suspect abscesses drained and their contents checked by a qualified vet.

who is willing to come by and help you when the chips are down. Locate a mentor (or two or three) via goat clubs, directories, and e-mail groups, and establish a working relationship before you need help (see the Resources section at the back of the book).

Assemble a caprine medical library and a well-stocked first aid kit. See the Resources section to read about our favorite veterinary manuals, and cruise government, university extension, and private goat Web sites for useful material, which you can print out. Arrange your printouts by subject matter, and file them in handy three-ring binders. Laminate important pieces such as kidding diagrams and keep them in your first aid or kidding kit. Still another option: download PDF files and store them on Zip disk or CD-ROM so you can access vital information with just a few quick clicks of the mouse.

It's important to keep a first aid kit handy not only for those inevitable emergencies but also for day-to-day treatment of cuts, scrapes, and dings. Keep the kit where you can readily find it, and replace each item as it's used. Three or four times a year, thoroughly inspect your first aid kit and discard any expired products.

Vaccinating

You can vaccinate for everything from caseous lymphadenitis to fibrosis, but don't stock up and systematically inoculate your herd. Goats don't need everything in the books. For example, when you unnecessarily inoculate using contagious ecthyma vaccine, goats shed their vaccination scabs and contaminate your

First Aid Kit

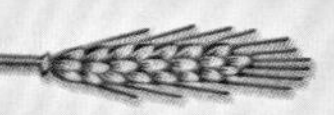

Meet with your vet and decide which emergencies you can face by yourself. If minor ones are all you care to tackle, a bare bones kit is enough. However, if you're an experienced livestock keeper or live far from your vet, he'll probably suggest prescription drugs to keep on hand. This is what we keep in our kit.

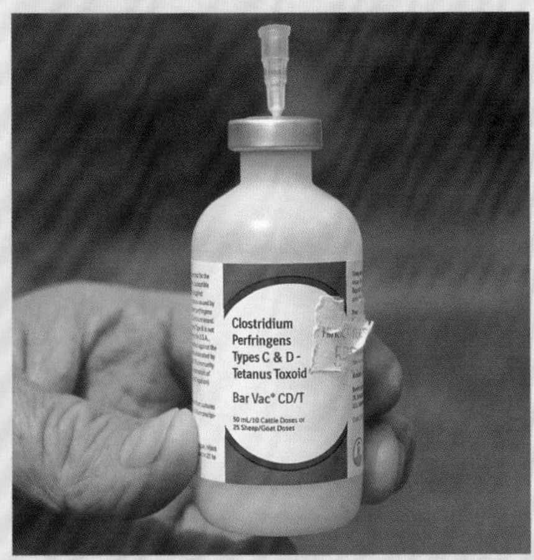

In the 'Fridge

Keep these items in the same box so they're easily accessible. If you have children, use a lockable tackle box.

Antibiotics

- Biomycin-200 (we use it in lieu of LA-200; they are equally effective, but LA-200 stings on injection)
- Penicillin
- Tylan-200

Vaccines

- C/D T toxoid
- C/D antitoxin
- Tetanus antitoxin
- Epinephrine for treating anaphylactic shock

Other Items

- Probios probiotic paste
- Goat Nutra-Drench
- Banamine
- Injectable vitamin B1
- Sterile water for flushing wounds

In the Emergency Bucket

We store our first aid kits in 5-gallon food-service buckets with snug lids. Since we also have a horse first aid kit stored in the same location, each is clearly marked with permanent marker. We use permanent marker as well to clearly indicate the contents of bagged items.

- A lead rope and halters in several sizes
- Scissors, several disposable scalpels, and a sharp folding knife (store together in plastic zipper bag)
- A powerful flashlight and extra batteries and bulbs (store together)
- Cotton-tipped swabs, stretch gauze, sterile pads, adhesive tape, two rolls of self-stick 4-inch bandage (we prefer Vetrap), and a digital thermometer (store together)
- Small containers of Kaopectate, milk of magnesia, Tagamet, and baby aspirin (store together)
- An assortment of 18- and 20-gauge, 3/4" disposable needles and 1-cc, 3-cc, and 6-cc disposable syringes (store together)
- Latex gloves
- Four-inch-wide duct tape
- Betadine scrub to clean wounds
- Schriener's Herbal Solution and emu oil, our favorite topical wound treatments
- Nitrofurazone salve to protect summertime wounds from fly-strike
- Blood-stop powder: never be without it!

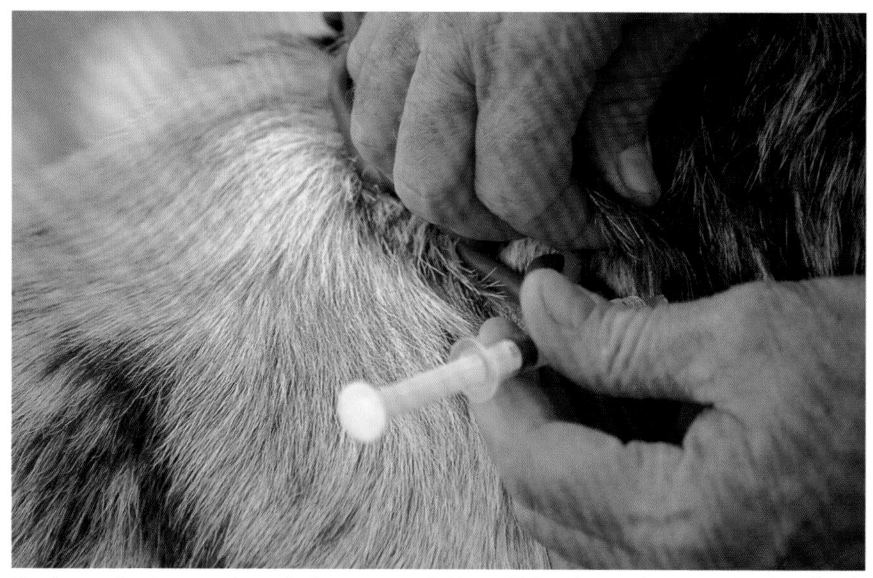

To give a subcutaneous shot, pinch up a tent of skin and inject the vaccine or medication directly under it, taking care not to poke through to the other side.

property with the virus, something you definitely don't want to happen!

Overvaccinating wastes money and stresses goats. Always vaccinate your goats with C/D T combination vaccine, an over-the-counter, combination product that protects them from *Clostridium perfringens* types C & D and tetanus. They need it no matter where you live. But discuss additional local and herd-specific needs with your vet or county extension agent before heading for the farm store's vaccine cooler.

Most goat keepers learn to inoculate their own goats. It's cost effective, and veterinarians are generally pleased to make fewer farm calls. But you must consult with your vet to formulate a vaccination program specific to your area and to your herd. Needs vary.

You can buy vaccines from your veterinarian, at many feed stores and farm stores, and by mail order from farm supply and biological warehouses. It's best to ask your veterinarian or an experienced goat breeder to show you how and where to give injections, even if you routinely vaccinate other farm animals or horses. See "Vaccinating Your Goats" box for tips.

Parasites

Like all other warm-blooded creatures (and some that aren't), goats are plagued by parasites. Some are external parasites and others are internal parasites. Below are descriptions of both.

Flies, Lice, Mites

In addition to everyday, in-your-face stable and biting horseflies, two specialized types of flies plague our goats: keds and botflies. Keds primarily infect sheep, but they do prey on goats as well. They're

Vaccinating Your Goats

Use disposable syringes and needles, and when you're through, dispose of them in a responsible manner.

Use a clean, new syringe for every session. It's best to use a new needle for each animal. Sharp needles cause less pain and work better. It pays to stay sharp (only about 30 cents a needle).

Choose 16- or 18-gauge needles in 1/2-, 5/8-, or 3/4-inch lengths. Longer needles easily bend or break. Shorter ones are perfect for giving subcutaneous (injected under a pinch of skin) shots, and goat vaccines are administered via that route.

Swirl a vaccine bottle's contents to mix it. Don't shake; you want to avoid making bubbles.

Pull back on the syringe's plunger a little farther than the volume of the shot you'll be giving. While holding the bottle upside down, poke the needle through the rubber stopper. Depress the plunger to inject air and avoid the creation of a vacuum. Pull back a little farther than the dose requires, then gently press the excess back into the bottle, removing any bubbles you may have created.

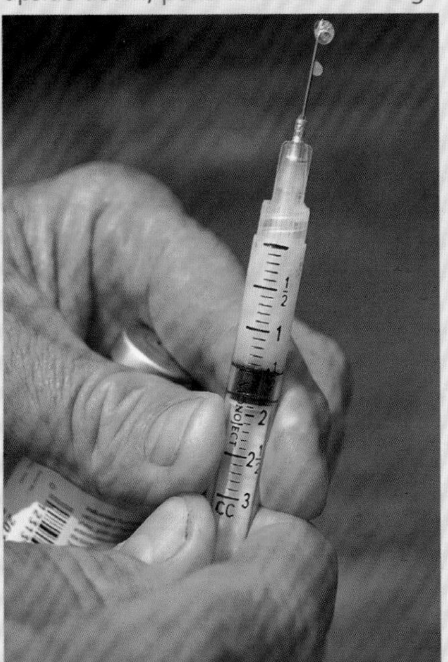

Always use a clean needle to withdraw vaccine from the bottle. A used needle contaminates the remaining contents. If you don't wish to use a new needle for each goat (though that's always best), insert a new needle into the bottle's rubber cap and leave it there. Attach your syringe to it to withdraw vaccine (as above), detach the filled syringe, and attach your used needle. Voilá, you're ready to shoot!

Give injections into clean, dry skin. Some vets recommend swabbing the area first with alcohol.

To give a subcutaneous injection, pinch up a fold of skin and slide the needle under it, parallel to the animal's body. Slowly depress the plunger, withdraw the needle, then rub the injection site to help distribute the vaccine. They can be given in the neck, over the ribs, or into the hairless area behind and below the armpit.

Intramuscular shots are trickier, but you'll rarely have to give one except to administer certain antibiotics. Ask someone to restrain the goat, then quickly but smoothly thrust the needle deep into muscle. The side of the neck is our preferred injection site. Always aspirate (pull back on the plunger about 1/4 inch) before you inject the contents. If blood sucks into the syringe, the needle pierced a vein. You must pull it out and try again.

Store leftover vaccines and antibiotics in your refrigerator, following the instructions on their labels. Discard leftovers after their expiration dates pass.

John gives Morgan a dose of ivermectin paste dewormer. Morgan isn't certain he approves.

wrinkly, brown, wingless flies that look like ticks and feed on blood. Botflies are fuzzy, yellowish-brown insects that resemble honeybees. They hover around a goat's nostrils where they deposit newly hatched larvae. The larvae migrate up the goat's nasal passages, feeding on mucus, until they reach the goat's sinuses. This naturally annoys the goat and can trigger severe inflammation and bacterial infections. After a time, larvae work their way back down the nasal passages, drop to the ground, pupate, and emerge as adult flies. Two generations in a single summer are not uncommon.

Several sorts of lice also live on goats. Some feed on skin and hair; others suck blood. Louse infestations cause extreme itchiness, skin irritations, rough coats, and hair loss. Lice are species-specific: goats can't pick up lice from poultry or birds.

Mites burrow into skin or feed on its surface, creating a fluid discharge and scaly, inflamed, denuded patches of skin called mange, scabies, or scab. Infestations are highly contagious and require aggressive treatment. One type, psoroptic mange (scabies), is a federal quarantine disease, so if you suspect your goats are infected, contact your vet without delay!

Worms

Goats are extremely susceptible to stomach and intestinal parasite infestation. Chronically wormy goats are scrawny, rough-coated, depressed, and anemic. They frequently suffer from diarrhea and usually die. Goats must be dewormed using the correct anathematics (worm-killing products) for the type or types of worms involved. Choosing dewormers at random simply won't work. No one dewormer kills every type of worm. You're throwing money away and endangering your goats by not using the right product for your herd's precise needs. Deworming is a complex subject to be discussed with your vet and your county extension agent, but in the meanwhile, keep the following points in mind.

The only way to know which vermicide to use is to take fresh manure samples for fecal analysis. When collecting samples, follow your goats with a sandwich baggie and collect "berries" as they fall (preferably, select nuggets that didn't come in direct contact with the ground). Take the samples to the vet, who will prepare smears and examine them under the microscope, searching for worm eggs. Depending on types and quantities, he can recommend a product based on your herd's exact needs. Or if you choose, buy a microscope and a parasitology reference book and learn to run fecals yourself.

Goats must be dosed according to weight. Underdosing is ineffective and leads to chemical resistance; overdosing can, depending on the product, kill your goats. So make sure you know your goat's weight. Though using livestock scales works best, they're prohibitively expensive for most hobby farm breeders. With a little effort, you can use a standard bathroom scale to weigh small goats. First weigh yourself, then pick up the goat and step back onto the scale; subtract your individual weight from the combined weight of you and the goat, and you have the animal's individual weight.

After weighing himself, John steps back onto the scale, goat in arms. Afterward, we'll do the math and have an accurate weight for our little goat.

Before using any dewormer, read the instructions. Not all are labeled for goats, so if your vet recommends an off-label dewormer, ask about cautions and restrictions. Follow instructions exactly.

Finally, if you choose homeopathic, herbal, or other organic dewormers, have fecals run on an ongoing basis. These products work well in some instances and fail miserably in others. Don't assume they're doing the job; for your goats' sake and yours, be certain.

The milk from newly dewormed dairy goats must not be used for human con-

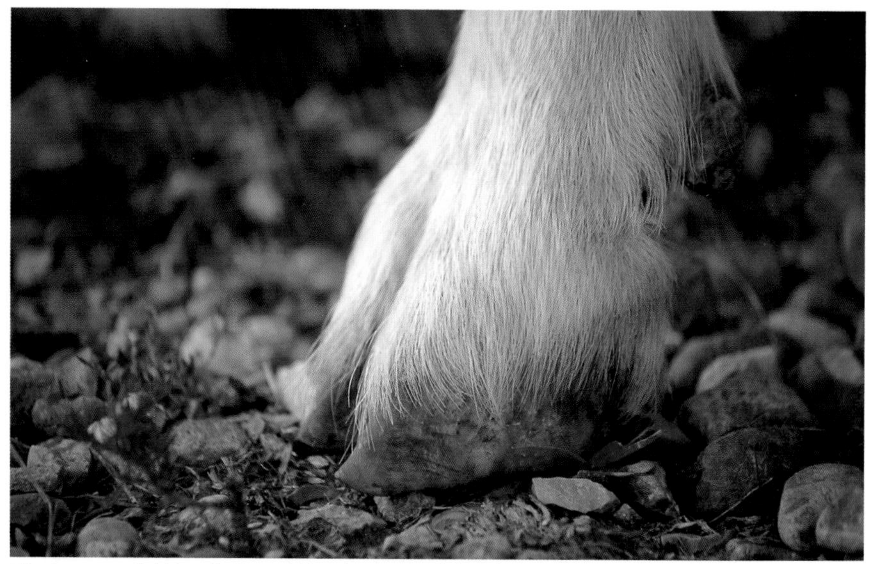

The bottom of this well-trimmed hoof runs parallel to the coronary band.

sumption; withdrawal periods vary from product to product. Ask your vet, or look it up online.

Hooves

Soil moisture and type, time of year, and breed influences how fast hooves grow. Trimming protects the integrity of your goats' hooves. In general, plan on giving pedicures at least two or three times a year, timing them to coincide with other labor-intensive procedures, such as worming and vaccinating. However, avoid trimming hooves during high-stress intervals such as extreme weather conditions, late pregnancy, or at weaning time. Hooves are easier to trim when they're moist.

Trimming

You'll need proper tools for hoof trimming. Most folks use standard hoof shears, but trimming with horse hoof nippers, a hoof knife, and a rasp works well, too. It's a matter of taste, experience, and convenience: people tend to use familiar tools they have on hand.

Safely tie your goat to a secure object using a sturdy lead rope attached to her halter or collar. Squat beside the goat, perch on an overturned bucket, or stand and lean over to trim.

Start trimming at the heel and work forward. Trim the heel even with the frog (the soft, central portion of each toe), then trim the walls level to match. If the frog is especially ragged, you can touch it up with a knife, taking paper-thin slices until you reach a hint of pink; the frog is a sensitive structure, so go no farther. When you're finished, the hoof should be flat on the bottom and parallel to the coronary band (the area where hoof and hair intersect).

When trimming a goat who has foot disease, trim her infected hooves last in

With an inexpensive set of hoof clippers, John has no trouble trimming our goats' hooves.

order to prevent spreading disease to healthy tissue. When you're finished, disinfect your tools to prevent infecting your other goats.

Not every lame goat has hoofrot. To evaluate a gimpy goat, watch her from afar. Which foot or feet is she favoring? How badly is she limping? Scan for foreign objects lodged in or between toes. Then carefully trim all four hooves. As you do, watch for signs of disease.

Dealing with Foot Rot

Foot rot and foot scald are closely related. In fact, they share a causative agent, the bacterium *Fusobacterium necrophorum.*

F. necrophorum is a common, hardy bacterium that dwells in soil and manure found on virtually every farm where livestock is kept. It causes thrush in horses and contributes to foot rot in cattle. It's an anaerobic organism (which means it can grow only in the absence of oxygen), so when animals are kept in dry, sanitary conditions, *F. necrophorum* poses no threat to them.

However, when hooves are continually immersed in warm mud and muck, bacteria invade the foot, often via a minor scratch or ding, causing foot scald, a moist, raw infection of the tissue between the sufferer's toes. Foot scald usually affects only one of the front feet. It's nasty and painful, and it frequently leads to full-blown foot rot.

Foot rot occurs when *F. necrophorum* is joined by *Bacteroides nodusus,*

Advice from the Farm

The Best Medicine

Those in the know talk about a few treatments for your ailing goat.

Magic Syrup

"Like lots of goat owners, we swear by Magic (a homemade blend of 1 part molasses, 1 part corn oil, 2 parts Karo syrup). We use it to put final finish on an animal or provide quick weight gain to get ready for a show. We give a 50-pound plus animal 50–60 cc every other day or so. It works, we use it often—it is part of our weight gain routine.

"We also use it on does in the early stages of toxemia, anemic goats, and goats who are sick and not eating as a support treatment to keep them eating and drinking. Vets here in Texas who have been doing sheep and goat research for the last thirty years think it is a great thing to give them when they are down or off feed. It's a really handy thing to have around."

—Robin L. Walters

Seeds for Prevention

"A handful of German black sunflower seeds added to your goats' diet each day will prevent goat polio. These seeds are high in thiamine, the roughage is good for the rumen, and the oils in the seeds are good for the coat."

—Bobbie Milsom

Sealer for a Sore Head

"Goats can break their horns, especially bucks who like to bash into things or aggressively head-butt their pen mates. If a horn is broken completely off and it isn't hemorrhaging or badly contaminated with dirt and debris, I've had good results from simply spraying the area with an aerosol antiseptic/sealer such as BluKote, keeping the animal isolated so that others will not bump the very sore head, and watching for any signs of infection.

"If the horn is broken at or near the base but is still reasonably well attached, the same approach works. As long as the soft tissue core of the horn is alive, new horn will grow and mend the break."

—Melody Hale

A Spark for Life

"Getting them in the house is a thing with me. I always bring them in and a lot of times lay them on the bed with me. It seems to give them that little extra spark, that small will to live. I don't have one shred of evidence or scientific support to back that up, but I know that it works a lot of times. When there's something here that is in dire straits, I spend every minute with that animal. When they open their eyes, they see me or feel my hand caressing them. It does matter to them."

—Donna Haas

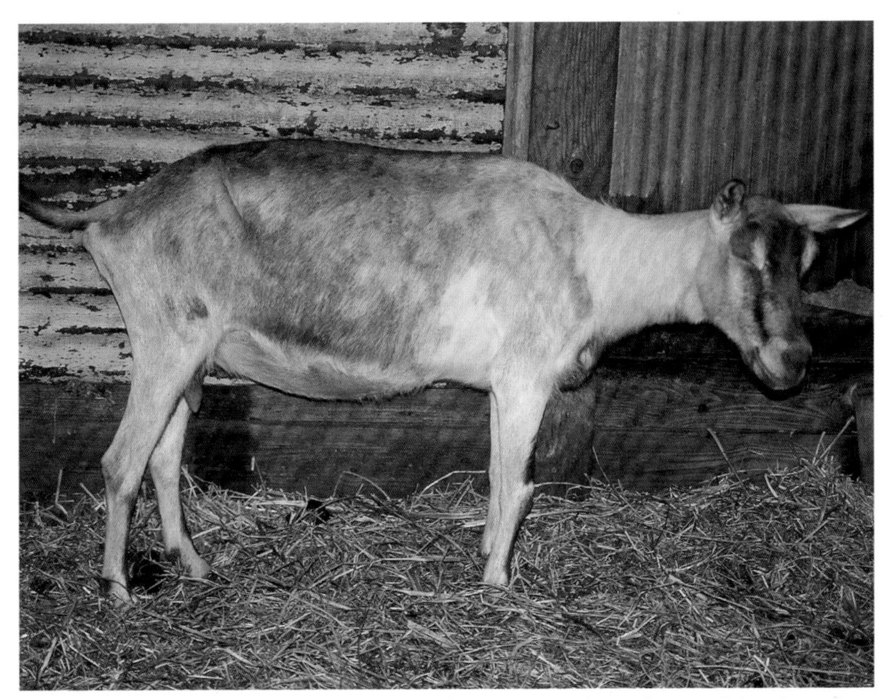

Well cared for goats age gracefully, as evidenced by this lovely old Anvil Acres Alpine, dam of our weanling wethers, Atticus and Arch.

another anaerobic bacterium that thrives in the hooves of domestic goats. It gains access via foot scald lesions and other injuries. When *F. necrophorum* is present, *B. nodusus* sets up house in the deeper layers of the skin, where it produces an enzyme that liquefies the tissue that surrounds it.

You can't miss foot rot; affected goats are very lame. Infected tissue is sleazy, slimy, and stinky. Infection beneath the wall and sole of the hoof causes the horny walls to partially detach. More than one hoof may be involved.

Foot rot is treatable, but it's a long, costly, time-intensive process, and in most herds, not an entirely successful one. The key to foot rot control is don't bring it home in the first place. If your goats don't already have it, they can't get it without coming into contact with *B. nodusus* bacteria.

To prevent its introduction, trim new goats' hooves on arrival, and quarantine them well away from your main herd for at least three weeks. Do the same with returning 4-H and show goats, goats who have boarded at your veterinarian's facility, or any other goat who leaves your farm and returns.

We can't say it enough: don't buy goats at livestock sales. Many producers knowingly dump infected stock at farm auctions. Even if the one you buy isn't infected, she's probably been exposed to infected goats and held in pens where *B. nodusus* thrives.

CHAPTER SIX

Bringing Kids into the World

Whether you breed to get your dairy does in milk, you want to bottle raise a pet or packing wether, or you raise kids for the ethnic meat market, learn all you can about the breeding process before you begin.

Choosing Breeding Stock

If you buy a $30 scrub goat at a farm sale and breed her to your neighbor's mixed breed buck, you'll get a single cute, $30 scrub kid. If your aim is to clear the woods of brambles and saplings, that may be enough. But if you want to raise quality animals, start with the best foundation stock you can afford.

Do your homework: know what types and breeds produce the kind of kids you want to raise. Review chapter 2, then skip ahead to the Resources section and visit Web sites listed under your fields of interest. Don't limit yourself. By breeding dairy or Angora does to high-quality meat breed bucks, many breeders produce marketable meat kids while pursuing their primary goals of producing mohair fiber or dairy products.

Don't buy a Pygora or Pygmy in a poke. Buy breeding stock from producers who keep detailed health, pedigree, production, and in the case of dairy breeds, milking records. Ask to see verification when buying from purportedly certified disease-free herds.

Choose mature animals who have already sired or produced quality kids. When choosing dairy goats, even bucks, try to see a prospective purchase's dam and, if possible, his or her sire's dam, too. Udder quality is highly hereditary—as is a tendency toward multiple birthing. Choose goats (especially bucks) from twin, triplet,

Boer kids mature quickly, reaching breeding age within a few months.

or quadruplet births to maximize their chances of also having multiple births.

Top-quality older goats can be best buys for entry-level and hobby farm goat producers. Large-scale breeders often cull at six to eight years of age, but with a little extra attention, goats in this age group have years of productivity ahead of them. They already know the ropes, especially at kidding time, a definite boon when their newbie owners don't.

Sex in the Goat Herd

Goats reach puberty at surprisingly early ages. Five to six months is the norm for most full-size breeds, but two-and-a-half-month-old bucklings have successfully impregnated their dams and sisters. To prevent unplanned pregnancies, wether males not destined for breeding and separate bucklings from the rest of the herd by the time they're twelve weeks old.

Most responsible producers breed doelings at eight to twelve months old; dairy breeders often say to breed them when they reach 80 pounds. In either case, doelings should be well grown and healthy, and they should be bred to bucks who (at maturity) are of the same size or smaller.

Most dairy goats are seasonal breeders. Their breeding season is triggered by decreasing daylight and runs from roughly late August through January. Some breeds, especially those originating in warmer climates, such as the West African Pygmy, South African Boer, and New Zealand Kiko, breed year-round.

Does cycle (come in heat) every eighteen to twenty-one days and remain in heat from eight hours to three days, ovulation occurring near the end of that period. Each doe's heat cycles differ from those of her herd mates, but her

A big Boer buck surveys his lovely harem. Boers with massive hindquarters such as these are said to be "hog butted."

cycle generally follows a pattern. If Tinkerbelle comes in heat every nineteen to twenty days and stays in heat about forty-eight hours, unless illness or stress throws off her biological rhythms, you can count on her following this pattern most of her life. Goats don't experience menopause, which means a doe will continue cycling until she dies. Many does kid into their mid-teens, but geriatric does experience more pregnancy-related problems than do younger does, so it's best to retire them at ten to twelve years of age.

We discussed goat breeding behavior in chapter 4, but we didn't touch on one important question: Should your buck live in the herd with your does? If you milk your does and your buck is in rut, definitely not. Does kept with bucks tend to give strong, off-flavored milk. It's hard to market "bucky-tasting" product.

In addition, if you're establishing a CAE- or Johne's-free herd, you *must* be present to remove newborn kids before they can nurse their dams. Many other goat owners (myself included) simply want to be there when kids are born to assist if needed. If you don't know when each doe was bred, you could miss the big event; so you don't want a buck running with your herd.

However, when a buck lives with his ladies, he has time and opportunity to properly court them. He's more likely to breed them as they ovulate, so conception occurs. Does are more likely to conceive with 24-7 exposure to a buck, which is why pasture breeding is the norm in meat-goat production.

If your buck doesn't run in the herd, you'll "hand breed" him. That is, you'll lead him to the doe or vice versa and leave them in a pen together until the deed is done (several times, in fact).

You'll repeat this performance every day until she rejects his advances.

If you don't own a buck, you can breed to someone else's by paying board and a stud fee. Or consider artificial insemination (AI). In addition to avoiding the cost and hassle of maintaining your own males, with AI you can choose top-quality bucks who complement each of your does. Goats can be inseminated using fresh cooled or frozen semen, generally resulting in a 60–65 percent conception rate. Many large-scale breeders and AI companies, such as BIO-Genics, Ltd. (see the Resources section), offer semen and insemination services. Or check with your county extension agent or breed club for local contacts.

Once does settle (become pregnant), they'll stop coming in heat. A vet can confirm pregnancy via ultrasound, or you can simply assume noncycling does have conceived.

The Waiting Game

Approximately 145–155 days after their last breeding date, depending on their breed, age, and previous production record, your pregnant does usually birth one to five kids.

It's wise to dry off lactating dairy does (take them out of production and allow them to stop producing milk) two or three months before kidding. This gives them time to rest and recuperate before their kids arrive and a new milking cycle begins.

Five or six weeks before kidding, boost pregnant does' C/D T vaccinations, trim their hooves, and worm them. (Read the labels before introducing new products or foods to pregnant does. For example, Valbazen, an especially effective white wormer, triggers abortion in pregnant sheep and goats.) At the same time, begin supplementing the does' diets with concentrates, based on type, breed, and body condition. Consult your county extension agent or local mentor for specific advice.

If your property (or the land where your feed and hay are grown) is selenium deficient, give each doe a Bo-Se (selenium/vitamin E) injection at four or five weeks predelivery. If you don't know, ask your county extension agent or your vet.

If your does succumb to ketosis, it will happen during the month before or the month after kidding. Monitor does' weight, make certain they exercise, and keep treatment materials at hand.

At least ten days before the first doe's due date, assemble a kidding kit or update the one you already have (see "Build a Better Kidding Kit" box). If you use jugs (individual mothering pens) instead of allowing your does to kid out on pasture, clean and disinfect existing pens or set up new ones in a well-ventilated, draft-free area in a shed or barn. Allow 25–35 square feet for each doe and her kids. Where drafts might pose a problem, opt for solid wooden walls. Bed with dust-free material (sawdust can trigger respiratory problems in newborns), and fit each pen with an elevated waterer and feeder. Don't use 5-gallon food service buckets or other large water containers in which tiny kids can drown.

A week before a doe's due date, clip her udder, escutcheon (the area between her rear udder attachment and her privates), vulva, and tail, especially if you keep dairy, fiber, or other long-haired goats.

Throughout kidding time, keep your fingernails clipped short and filed in case you need to internally reposition a kid. Review your educational material, be it books or bulletins from the university sites in our Resources section, and know in advance how to recognize problems and correct them. Post your vet's number by the barn phone, and add it to your auto-dial.

Delivery Day

A week or so before the first expected delivery day, start monitoring those does! Does tend to exhibit the same set of prekidding signals from year to year, but each is an individual and no two follow exactly the same routine.

Most first-timers begin building an udder (their udders start developing) four to six weeks prior to kidding. The average veteran doe bags up (her udder begins filling with milk) beginning ten days before and continuing up to the very day she delivers. When delivery is imminent, does' udders are full and feel tight.

A doe's tail ligaments—the ones stretching from just above the spot where her tail joins her spine to her pin bones (those bony protrusions on either cheek of her butt)—become more elastic as delivery approaches. Check them twice a day. They'll change from hard to soft to mushy feeling. At the mushy stage, she's roughly twelve hours from delivery. (See the Resources section for Web sites with how-to photos on checking the ligaments.)

A day or two before kidding, many does drift away from the main group, sometimes taking along a grown daughter or a friend. A doe may seem introspective or preoccupied and wander around as though looking for something. During the same period, a long clear string of mucus may trail from the doe's vulva.

From a day to just minutes before giving birth, does usually begin nesting—pawing the ground, turning around, lying down, and getting up again, over and over and over. A doe may stretch a lot or yawn or even murmur to her in utero babies in a soft, subdued voice.

When a doe gets down to business and starts pushing hard, stay calm. Don't help unless she needs it, but be ready to act quickly and definitively if she does. Does can deliver standing or lying down; either is normal.

The first thing to appear at her vulva is a translucent bubble: the amnion, filled with amniotic fluid (this bubble can safely rupture at any time). As it emerges, you'll see first one and then another hoof, and eventually a little nose will appear. Once his shoulders are delivered, the kid usually plops right out.

Build a Better Kidding Kit

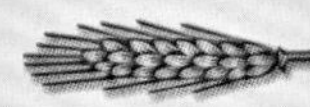

Kidding is the most rewarding part of the goat keeper's year. Usually the process goes without a hitch, but glitches can occur, so the goat keeper should assemble a kidding kit to field possible emergencies. Here's how we do it on our farm.

We pack our lambing and kidding supplies in two containers. The one we take to the barn is a hard plastic step stool with a storage compartment inside. It is sturdy and tip resistant and holds a lot of gear, and on cold, wet nights, it sure beats sitting on the ground. It contains:

• **Sharp scissors** to trim the umbilical cord to an inch or so in length. We disinfect them after each birthing and slip them in a plastic zipper bag (we use a lot of plastic zipper bags in our kit) to keep them clean.

• A **hemostat** to temporarily clamp on the umbilical cord if it continues bleeding (disinfected and kept with the scissors).

• **7 percent iodine** to dip the cord into after trimming. Some folks squirt iodine on the navel while the lamb or kid is lying down, but it's much cleaner to dip the navel into a shot glass while the baby is standing.

• **Dental floss** to tie off bleeding umbilical cords if needed.

• **Two flashlights**—we like packing a backup in case the first light malfunctions. We tuck them in another plastic zipper bag to keep them dry.

• **Lots of lubricant** for repositioning babies. We like Suberlube and keep two squeeze containers in our kit.

• **Betadine scrub** to swab a doe or ewe's vulva before repositioning babies.

• **Shoulder-length OB gloves**—sterile, individually packaged ones. They're harder to find than nonsterile gloves, but they're worth the search.

• A **sharp pocketknife**, so we don't have to use our umbilical cord scissors for routine cutting chores.

• A **digital thermometer**, the kind that beeps.

• A **bulb syringe** designed for human infants. It can't be beat for sucking mucus out of tiny nostrils.

• An **adjustable, rubber kid and lamb puller.** This, the thermometer, the knife, the gloves, and other small items are stowed together in a single plastic zipper bag. We'd add a paint stick designed for livestock or other ID marker to the mix if we had more animals, but our herd is so small that telling which babies go with which mom isn't a problem.

• **Nutri-Drench** (one labeled for goats) for weak newborns and exhausted moms, including a catheter-tip syringe with which to give the stuff.

• An **adjustable halter and lead**, with which it's easier to move most moms than it is without one.

• A **lamb and kid sling**— a back saver!

• **Towels**—soft, old, cleanup toweling is stuffed into any remaining space.

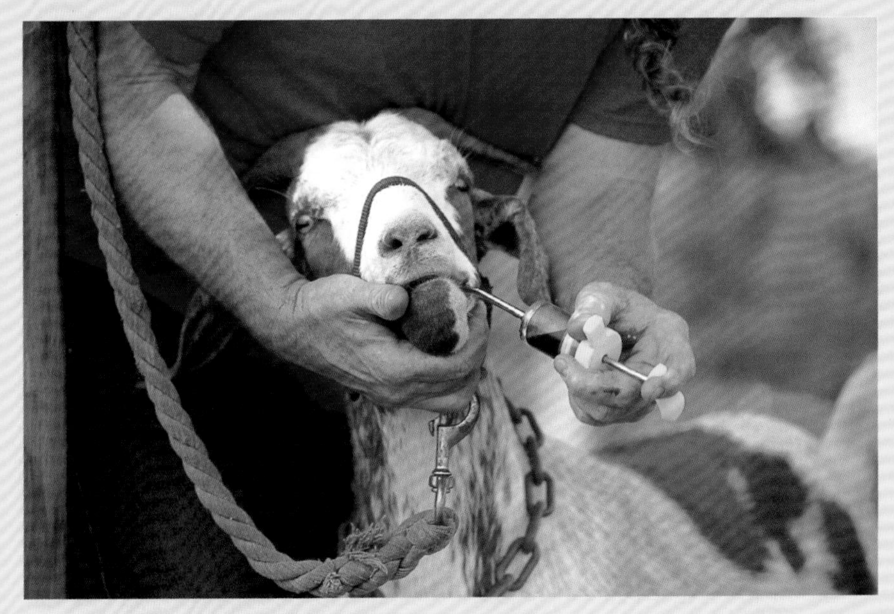

John gives Eamon several ounces of Goat Nutra-Drench from a standard dose syringe.

Having once scrubbed spilled lube out of a kidding kit, we make certain the iodine, the lube, and the Nutri-Drench are individually double-bagged.

Our other container is a lidded 5-gallon pail. It stays in the house, and its function is to keep our other lambing and kidding supplies centrally located. It houses the following items.

• **Milk replacers** (specifically for kids)—we repackage them in plastic zipper bags and store 3 to 4 pounds in the birthing supplies container, the rest in tightly sealed tins.

• A **plastic calibrated nursing bottle** with a **Pritchard teat** and several **spare teats**.

• A **flexible plastic feeding tube**, a **felt-tip marker**, and a **60-cc syringe** for tube-feeding weak newborns.

• A **16-ounce measuring cup**, so we don't have to dig through the cupboards to find one when we need it; it's also great to store the spare Pritchard teats in. It's a big one, so it can double as a milking receptacle.

• A **small whisk** for mixing milk replacer. All feeding supplies, including the measuring cup, are stored in a second plastic zipper bag.

• **Syringes and needles** go into another bag, along with an **elastrator** and the **rings**.

Other folks add different items to their kits. What you need depends somewhat on your goat's breed and on where you live.

Four little Boers snooze in the sun. Multiple births are the norm in most breeds.

Strip excess fluid from his nose by running your fingers from below his eyes to his nostrils, or use a human infant–type bulb syringe to suck it out. Place him in front of his dam so she can begin cleaning him. This is important: this is when she bonds with her kids.

If more babies are imminent, she'll repeat the process until they're all delivered. Never leave until you're sure the last has arrived! Move each one to the side immediately so he doesn't get stepped on. This is a good time to trim the kid's umbilical cord if it's more than 2 inches long. Hold a shot glass or similar container full of 7 percent iodine to the kid's navel area for several seconds. Make certain the cord is totally saturated, and use fresh iodine for each kid. *Don't omit this step.*

If the cord doesn't stop bleeding, apply a commercial navel clamp about an inch below the kid's belly, or tie the cord's end off with clean dental floss.

After her ordeal, your doe will be tired and thirsty. Bring her a bucket of lukewarm water (perhaps adding a dollop of molasses as a pick-me-up), and give her a nice feed of hay. It's important to leave the new mom alone with her kids so that they can bond, but you have a few more tasks before you go.

Caring for the New Kids

Make sure your kids get the right start in life so they'll be healthy. That means ensuring they get the proper nutrients, keep warm, and are protected from disease. Bottle-fed kids have their own special needs. You must also make a decision about whether to castrate and disbud and when.

Food, Shelter, Health

When the babies have arrived, milk a stream of fluid from each of the doe's teats to clear any wax plugs. This first milk, a thick, yellowish fluid called

Training a Bottle Kid

Training a bottle kid is fun, yet frustrating. We use this method and feeding schedule developed by veteran California Red sheep breeder Lyn Brown. It works just as well for kids.

Sit on the floor with your legs crossed. Place the kid in your lap facing away from you, sitting on his butt with his legs straight out in front of him. Cup his jaw with your left hand, open his mouth and insert the nipple, then steady the nipple using the fingers of your left hand. This keeps the nipple aligned with his jaw and his head in a natural nursing position, essential to keep milk from spilling into his rumen. He'll do his best to avoid the nipple, but persevere.

Elevate the bottle just enough to keep the nipple filled with milk; as the bottle empties, you'll add more tilt. With your left hand on his throat, you can easily tell if he's swallowing. This is important—you want milk in his tummy, not in his lungs.

Make certain you don't overfeed. "Most people kill their first bottle baby with kindness," Brown explains. "They overfeed it because the baby cries and they think it must be hungry. I know I did. Now I follow this feeding schedule, no exceptions. If our lambs [kids in this case] cry between feeds, we give them Pedialyte or Gatorade. That won't hurt them as far as enterotoxemia goes while filling the void for them." The following amounts are calculated for full-size sheep or medium-size goats.

- Days 1–2: two to three oz, 6x/day (colostrum or formula with colostrum replacer powder)
- Days 3–4: three to five oz, 6x/day (gradually changing over to species-specific milk replacer)
- Days 5–14: four to six oz, 4x/day
- Days 15–21: six to eight oz, 4x/day
- Days 22–35: gradually work up to sixteen oz, 3x/day

"At about 6 weeks," Brown continues, "I begin slowly decreasing the morning and evening feedings and leave the middle feeding 16 oz, until I eliminate the morning and evening bottle entirely (remember, they are eating their share of hay or pasture by now). I continue with the one 16-oz bottle for about two weeks, then eliminate the bottle feedings entirely."

A Boergora (half Boer, half Angora) doe feeds her twins, who nurse from a kneeling position with their heads thrown back. This causes a band of tissue in the esophagus to close, allowing milk to bypass the nonfunctioning rumen and flow directly into the abomasum.

colostrum, is packed with nutrients and antibodies essential to the kids' survival. The antibodies in colostrum are present for only about twenty-four hours after kidding. A newborn should ingest his first meal of colostrum within two hours; every kid should nurse before you leave. If kids don't nurse on their own, milk the doe and bottle- or tube-feed them the first meal. Once they've tasted this elixir of life, most kids will eagerly seek the lunch bar for themselves.

Young kids must be kept reasonably warm. Some people install heat lamps above jugs and bottle kids' pens, but because these lamps often cause barn fires, their use is risky. A solid-sided jug or pen in a draft-free section of the barn, when deeply bedded with long-stem hay or straw, is warm enough in all but the coldest climates. A comfy, well-bedded doghouse or airline-style dog crate with an old blanket or two draped over the top makes a dandy addition to bottle babies' quarters. Or install a bottle kid pen in your home. Lift-top wire dog crates and puppy exercise pens with tarps spread beneath them make fine kid housing. Bedded with old, frequently laundered blankets, indoor kids produce little odor. They can even don human diapers and frolic through the house. Our boys spent their infancy in our living room!

Kids are susceptible to conditions as diverse as constipation and scours, pneumonia, acidosis, enterotoxemia, floppy kid syndrome, coccidiosis, tetanus, goat polio, and white muscle disease. Learn all you can about these problems before kidding time; the Resources and our Appendix will point the way. Weak kids

Bottle-Feeding Equipment

You'll need proper equipment to feed your kids. We prefer the Pritchard teat: an oddly shaped, soft, red nipple with a yellow plastic base incorporating a flutter valve to regulate airflow. Pritchard teats can be screwed onto calibrated lamb and kid feeding bottles or onto any type of household bottle with a 28mm neck (20-ounce plastic soda bottles are ideal). The first time we use a soda bottle, we measure out an individual feeding, pour it into the bottle, then mark the fluid level with a felt-tip permanent marker. We use a bottle for a few days, toss it into the recyclables, and substitute a new one. Easy-flow human infant bottles and nipples also work well, as do the other soft nipples sold by goat and sheep supply outlets.

You'll also need a measuring cup and, if you are feeding milk replacer, a mixing bowl and a wire whisk. Keep all feeding supplies squeaky clean! Wash and rinse them after each use, and once a day, briefly soak everything except the nipples in a weak bleach solution (1 part bleach, 10 parts water).

Morgan chugs his noon feeding from a 20-oz plastic soda bottle fitted with a Pritchard teat.

Eight-year-old Alyssa Rockers of Carthage, Missouri, cuddles bottle kids while visiting MAC Goats.

must be tube-fed until they're strong enough to stand and suckle. This sounds scarier than it is. Ask your vet or mentor to show you how to pass a stomach tube and have one ready in case you need it.

Bottle Kids

Unless they are fostered on another willing doe, orphan and rejected kids must be bottle-fed—and what's more fun than raising a bottle kid or two? To bottle kids, you're Mom, herd queen, and best friend all rolled into one. They will carry that attitude into adulthood. In fact, recreational goat owners routinely bottle-feed for precisely this reason.

Many producers haven't time to bottle-feed orphans and rejects, so they give away or sell the kids cheaply to those who do. If you'd like one, put out the word. Contact local breeders, and post to regional goat-oriented e-mail lists. Tell local vets and your county agent. You're bound to find a likely kid or two.

If you have room, two kids are better than one. They'll entertain one another when you're gone, and if you feed milk replacer, you'll save by buying in larger volume. Kids can be fed individually or in groups, using a rack-type bottle holder or multiple-nipple feeder.

The kids you accept should have fed on colostrum for at least the first twenty-four hours of their lives. If you're called to come pick up newborns and they haven't gotten any, ask if you can buy colostrum from the breeder. If fresh or frozen goat colostrum isn't available, cow or sheep colostrum will do. Other alternatives include CL-Nanny Replacer Colostrum or CL-Ewe Replacer Colostrum (available from Mid-States Wool Growers Cooperative Association) or Goat Serum Concentrate fed with Goat Colostrum Replacer (sold by Hoegger Goat Supply; both are listed in the Resources section). Don't rely on colostrum "boosters" of any sort; they

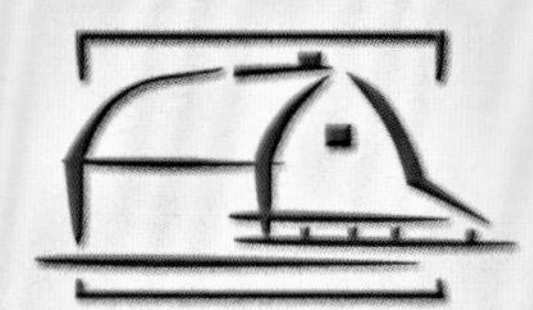

Advice from the Farm

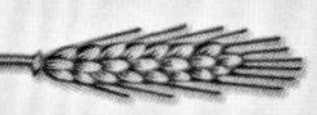

Kidding Time

Kidding tips from our panel of experts.

All Signs Are a Go

"The doe should be sunk in by the tail head, and her udder should be a shiny pinkish. Sometimes the udder doesn't engorge before labor, but usually it's huge. There will be a vaginal discharge and the doe will dig at the ground, lie down, and strain. If she's not doing any of these, she's not in labor. If the kid is laying wrong, the doe doesn't go into hard labor, but you'll see signs."

—Pat Smith

Out of His System

"Once a kid has gotten the black meconium (first manure) out of its system and has had enough milk make it through, then the poop will appear yellow. If it makes a blob that sticks, then it has to be removed or the kid won't be able to poop and that will cause severe infection and death.

"Just slip on a pair of those disposable gloves and carefully pull it off. You may have to wet it down to soften it enough to release. Once it's off, you can apply Vaseline or something that will keep it from sticking again."

—Rikke D. Giles

Bringing Up Baby

"We have our barn (keep in mind it used to be a stable) arranged so that every stall has a baby access door to the center runway. That gives kids an escape from mom or a doe from hell and gets them socialized with the other kids they will be penning with later. We also have baby-height feeders in the runway so they can eat without having to fight to get to the food. It also gets them used to seeing and being around me. A good 90 percent of my kids are friendly enough to pet all the time; the other 10 percent of them are usually the ones who prefer to stay close to mama."

—Rikke D. Giles

Most goat producers who castrate young goats do so with an elastrator tool and elastrator rings such as these.

simply aren't enough to do the trick. If your kids don't ingest real colostrum or a viable alternative during those critical first twenty-four hours (forty-eight hours are better), they'll lack vital immunity to disease. Many such kids do survive, but you must be especially vigilant and get them to a vet at the first sign of illness because they won't have the wherewithal to fight it on their own.

What do you feed bottle kids after colostrum? Goat's milk always works best, but these mixtures using store-bought cow's milk work well, too:

- 1 part dairy half-and-half; 5 parts whole milk
- 1 gallon whole milk; 1 cup buttermilk; 12 ounces of evaporated milk

If you use commercial powdered milk replacers, *always* buy high-quality milk-based products designed specifically for kids. Soy-based replacers, products designed for the young of other species, and one-type-fits-all-species milk replacers absolutely *will not* do. If you use a milk replacer, read the label carefully and measure ingredients every time. Slap-dash mixing leads to potentially serious upsets such as bloat, diarrhea, and enterotoxemia. Mix only enough replacer for a day. Keep it in the refrigerator, and don't return unused portions to the jug. Buy enough of the same brand to last your kids through weaning, as switching products leads to gastric upsets. If you must switch, do it gradually over the course of at least ten days.

Castrating and Disbudding

If you castrate male kids, the easiest and least expensive way to do it is to band them before they're two weeks old. A pliers-like tool called an elastrator is used to stretch thick, strong, rubber bands wide enough to slip over a kid's testicles.

Breeders may choose to leave horns intact, but if the goat will be handled much, it's safer to remove them.

Advice from the Farm

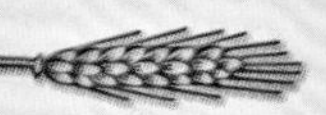

Bottle Brigade

Tips for the bottle brigade from those who have been there.

Not Too Much

"There is nothing wrong with feeding kids milk replacer, just make sure that you find one designed for goats and made out of milk protein and not out of soy. However, milk replacer seems to be more unforgiving than real goat's milk.

"I am not going to waste my money on feeding goats store-bought milk, so I try to keep a dairy goat fresh, and milk in the freezer; goat's milk is a lot easier than mixing up replacer.

"Make sure you stick to a schedule when bottle feeding. An over-hungry kid is going to eat too fast. In general, bottle kids, given a chance, drink too much and too fast, and this can lead to bloat. In order to help prevent bloat and other digestive problems, I give my bottle kids C/D antitoxin and Poly-Serum every three weeks; it works."

—Robin L. Walters

Not Too Hot

"Whatever kind of milk you are feeding, be sure that it's warm enough, but not too hot. Body temperature. Use a soft nipple because kids don't like hard ones. Force the nipple in the kid's mouth and hold it until he drinks. Be careful you don't drown him. He'll catch on."

—Pat Smith

To Bottle or Not?

"Kids don't have to be bottle raised to be friendly. I have a 300-pound Boer buck that was dam raised and pasture bred. When I got him at eighteen months, the only time anyone had put a hand on him was to worm or vaccinate him. He is a big baby, and as sweet as can be. He'll even come up to me and duck his head for me to scratch him."

—Robin L. Walters

Salem scrubs his forehead on a water tub. Kids rub on everything when their horns are emerging.

Lack of circulation causes the scrotum to wither and slough off in four to eight weeks. If you use this method, give each kid a shot of tetanus antitoxin (not tetanus toxoid) when you band him; *never omit this step!* However, since penis development ceases when a kid is castrated, authorities believe early-banded wethers are more likely to develop urinary calculi than are late-castrates and bucks. Many recreational and pet owners delay the procedure until kids are at least three months old, then have a veterinarian castrate them under sedation and local anesthesia. Meat producers selling 40–60 pound kids to ethnic markets needn't castrate at all. In fact, some ethnic communities pay premium prices for bucklings.

If you don't want horned goats, kids should be disbudded when they're three to fourteen days old. Disbudding is accomplished by destroying emerging horn buds with a red-hot iron; it's not a job for the squeamish or uninformed. Ask a vet or your goat mentor to show you how it's done, then buy the proper tools before doing the job yourself. Or have your vet disbud kids using a local anesthetic.

CHAPTER SEVEN

Making Money with Goats

There are three avenues for making money with your goats: meat, dairy, and fiber. Each has advantages and disadvantages, so carefully weigh all the factors before deciding which avenue will work best for you.

Meat Goats: The Mortgage Lifters

Farmers and ranchers across the United States and Canada are calling meat goats "the mortgage lifters" of the new millennium. Easy-to-handle, easy-to-raise meat goats are selling at all-time high prices, yet more goats and more goat producers are needed to supply North America's burgeoning goat meat market with market goats and the breeding stock needed to produce them. Experts have characterized meat goat production as the fastest growing animal enterprise in the country.

Thirty years ago, few Americans ate goat meat. Most North American goats were dairy goats, Angora goats sheared for their mohair fleece, and Spanish scrub goats used for brush control. The scrub goats were sometimes used for meat, most of which was sold on the hoof and shipped to Mexico. New U.S. Department of Agriculture (USDA) statistics tell a completely different story. According to the USDA National Agricultural Statistics Service's first annual goat survey, on January 1, 2005, the U.S. goat inventory totaled 2.5 million. Breeding goats totaled 2.1 million and market goats 0.4 million. Milk goats numbered 283,500 head, Angora goats 274,000, meat and all other goats 1,970,000.

Consider this: the USDA estimates that roughly 50–60 percent of America's meat goats are processed at USDA-inspected abattoirs. The first year goat slaughter statistics were kept was in 1977, when 35,000 goats were tallied. By 1985, the figure

Commercial meat goat producers routinely breed Spanish does (like this handsome white one) and Boer and Kiko bucks to produce meaty, highly marketable kids.

had grown to 124,000. Only eight years later, 320,000 goats were accounted for, amounting to more than a 900 percent growth in only sixteen years. In 2000, federally inspected slaughterhouses processed 549,000 goats.

Why the Demand?

The reason for this unprecedented industry growth is simple: 65–70 percent of all red meat consumed globally is goat meat, and America's expanding ethnic population is willing to pay premium prices to buy it. Families of Mediterranean, southern European, Middle Eastern, African, Southeast Asian, South American, Central American, and the West Indies extraction all favor goat.

Cabrito (the flesh of 10–15 pound milk-fed kids) and *chevon* (the meat of older kids and mature goats) are favorite fare in Hispanic households. The U.S. Census Bureau projects that between 1995 and 2050, Hispanics will account for 57 percent of the immigration into the United States and that Hispanics will account for 25 percent of the U.S. population by 2050.

Kid goat is the traditional mainstay of Muslim feasts served before Ramadan, at 'Id al-Fitr and at 'Id al-Adha. The Muslim population in the United States, though not a large percentage of the population, is a significant and growing segment. The Muslim converts of non–Middle Eastern origin compose a substantial fraction of the total, and they are said to be particularly observant of traditional Muslim dietary preferences.

A second, smaller segment of America's population is clamoring for

The Healthy Meat

Goat Meat Versus Other Meats

Meat (per 3 oz roasted)	Calories	Grams of Total Fat	Grams of Protein
Goat	*122*	*2.5*	*23*
Beef	245	16	23
Chicken	120	3.5	21
Lamb	235	16	22
Pork	310	24	21

goat meat. Health-conscious individuals turn to naturally lean goat meat for its health-giving qualities. Compared with beef, pork, and lamb, it's lower in calories and fat and equal or higher in protein. Although chicken is slightly lower in calories than goat, it is also lower in protein and higher in fat. People are also discovering just how great goat meat tastes.

Why Are Farmers Flocking to Goats?

At present, goat producers cannot supply enough market goats to meet North America's skyrocketing demand for goat meat. An astounding amount of product has to be imported. For instance, in 2003–2004, Australia exported 16,097 metric tons of goat meat, 48.6 percent of it to the United States. The value of this goat meat was slightly over $28 million Australian (approximately, $21.5 million U.S. dollars). The fact that America's demand for goat meat far exceeds the domestic supply means a ready, established market for new goat producers.

Other appealing aspects: goat producers can choose from a number of viable business options, depending on what best fits their interests and circumstances. Start-up costs are relatively low, business can be expanded rapidly, and land is not usually a problem.

Finally, goats are intelligent, friendly, and just plain fun to have around. Except for the occasional ornery buck, they're nonaggressive and easy to handle, even by children and seniors. Almost without exception, the goat producers I've talked to say the best part of farming goats is the goats.

Business Options

Most commercial producers maintain large herds of unregistered and crossbred goats. Their objective is to produce fast-maturing, low-cost kids for slaughter.

A junior showman confidently displays his equally young goat. General ease of handling, even by youngsters, is one of the reasons farmers like keeping goats.

Commercial producers market live meat goats (usually by the pound) directly from their farms, through livestock auctions, or to buyers and brokers. Slaughter goat prices today range $0.65–$2.00+ per pound depending on age, grade, and availability.

Registered show goat breeders maintain fewer, but far more costly, registered animals. Their goal is to use popular genetics to produce goats capable of winning in stiff competition at major goat shows. Goats in this group are each currently selling in the $1,500–$20,000 price range.

One group of breeding stock producers markets high-quality, fast-maturing registered goats of popular bloodlines to other breeders whose goal is herd improvement, rather than show ring victories. Such goats currently fetch prices in the $500–$2,000 range, with quality bucks selling higher than does.

A second group produces quality percentage (partbred) stock of the popular breeds—usually Boer or Kiko—by breeding top-flight bucks to lower percentage does. They market does and bucks to other breeders, high-quality wether kids as 4-H/FFA show stock, and low-quality and excess bucks for slaughter. Percentage breeding stock currently costs $150–$500; 4-H/FFA wethers, $90–$500.

Many goat dairies, large and small, are entering the commercial meat mar-

Some breeders like the beauty of paint Boers such as this rich red paint MAC Goat doeling.

ket by breeding their milking does to purebred or high percentage Boer or Kiko bucks. Meat kids are raised on excess goat's milk or on milk replacer, while their mothers continue working the milk line.

Costs, Expansion, Land

Compared with other livestock ventures, start-up costs are unusually low, especially for entry-level commercial meat goat producers. Good quality commercial breeding stock is inexpensive and readily available. Although new goat fencing can be costly, in most cases existing fencing, housing, and handling facilities are easily and inexpensively converted for goats.

It's possible to start small and expand rapidly by retaining doelings and marketing only male kids for slaughter. Routine multiple births (two kids are the norm, but up to four not uncommon) equate with rapid opportunities for herd expansion.

It doesn't take a lot of land to raise goats. Many registered breeding stock producers—who don't need a lot of goats to show a handsome profit—operate from small farms.

Pastured goats, cattle, and horses prefer different plant species and can be pastured together or in rotational grazing programs. Measured in amount of lean product per unit of input, goats maintained on lush pasture or in a feedlot scenario can't compete with cattle, sheep, or hogs. However, pastured on brushy, weedy, rocky browse, goats top the others hands down. A further advantage: seven to eight goats flourish on the dry feed that a single beef cow would consume.

Goats prefer rough browse. With minimal supplementation, they produce marketable kids on land that would

A group of Toggenburg goats and a bay horse make congenial pasture mates on this small farm in Wisconsin.

starve beef cattle. They clear land of brush, kudzu, leafy spurge, purple knapweed, wild blackberry, and multiflora rose, producing marketable meat while improving woodlots and destroying noxious weeds.

With goats, location barely matters. Meat goats can be marketed live from the farm, at livestock auctions, or to brokers who truck large numbers of live goats to goat slaughtering facilities. Although the majority of federally inspected goat-slaughtering facilities are located in Pennsylvania, Missouri, Texas, Delaware-Maryland, and Illinois, in most areas producers have formed marketing organizations to cooperatively ship their own goats to slaughter, thus eliminating the middleman and earning additional profits for themselves.

Starting Right with Meat Goats

Don't rush blindly into raising meat goats. Settle on a niche (commercial meat kids? show wethers? registered breeding stock?), then do your homework. You *must* have a viable market for your product, no matter what type of meat goat enterprise you choose.

Talk with your county extension agent. Ask about local marketing opportunities. Contact others in your region who are successfully engaged in the business you choose. Tell them what you have in mind and ask for their feedback. Ask lots of questions. They are your number one source for local goat business information.

Subscribe to meat goat periodicals; buy back issues if you can. Spend time

Commercial meat goats come in all shapes and colors.

online perusing sites and services listed in our Resources section; a tremendous amount of meat goat material is available for free on the Internet. Join e-mail lists and discussion groups such as chevontalk and BoerGoats in the Yahoo online network. Attend meat goat seminars sponsored by goat organizations and universities or private seminars such as Suzanne Gasparotto's Goat Camp throughout the United States (see Resources).

Meat goats are hot, hot, hot. If you can raise them economically and find a steady market, you're almost sure to show a profit. And it'll be a long time before supply exceeds demand.

Honing in on Ethnic Markets

If you want to become a commercial producer, whether you plan to sell direct from your farm, at livestock sales, or to livestock brokers, you'll need to raise kids who meet the needs of your buyers. For instance, Muslim buyers prefer lean, dressed goat carcasses in the 35–40 pound range and only consume halal meat (*halal* means "permitted"; in this case, meat from goats slaughtered according to Islamic law). Hispanic buyers buy cabrito or meat from older kids, and Caribbean buyers tend to prefer mature goats.

The following chart is adapted from a University of Illinois Extension bulletin, "MarketMaker Chicago Ethnic Markets: Goats" (http://www.marketmaker.uiuc.edu/PDF/ethnicgoat.pdf). If you plan to market goat meat to ethnic buyers, you'd be wise to download and read it. Once you've targeted an ethnic community, learn more about its goat meat needs via the Internet and other sources.

Ethnic Groups

Carcass Preference by Ethnic Group

Group	Preferences (dressed weights)	Comments
Asian (e.g., Chinese, Korean, Thai, Vietnamese)	High-quality 60–70 lb goats	Prefer a headless carcass, feet off, with scalded and scraped skin.
Greek	30–40 lb kids	
Hispanic	Milk-fed cabrito (5–12 lb), high-quality 15–25 lb kids	Sold to retailers head on, feet off, skinless; retailers sell ribs, whole legs, stew meat.
Italian	20–25 lb young kids	
Jewish	20–40 lb high-quality kids, newly weaned, milk-fed	
Muslim	Fairly lean, less than one year old; 35–40 lb	Goat and lamb are served at traditional holiday feasts; also served at special occasions such as weddings and birth celebrations; retailers sell whole, half, and quarter goats as well as various cuts; heads, tongues, liver, brains, and kidneys are sold as well.
West African/Caribbean	Lower-quality mature goats; intact males preferred	Caribbean outlets sell goat legs, ribcages, and stew meat; African groceries carry whole goat legs, quarters, shoulders, and stew meat. Some carry carcasses with the skin left on.

In the Han Ah Reum International Supermarket in Virginia, an Asian shopper considers a variety of meat selections.

It's also important to learn when your target group's religious holidays occur. For example:

Muslim Religious Observances*

Eid al-Adha	October 16, 2013 October 5, 2014 September 24, 2015 September 12, 2016	Fairly lean, unblemished kids (wethers are unacceptable) less than one year old, 35–40 lb dressed weight.
Ramadan (start)	June 29, 2014 June 18, 2015 June 6, 2016 May 27, 2017 May 16, 2018	Kids less than one year old, fairly lean, 33–40 lb dressed weight.
Eid al-Fitr	July 29, 2014 July 18, 2015 July 6, 2016 June 26, 2017 June 16, 2018	Kids less than one year old, fairly lean, 33–40 lb dressed weight.

* Dates are approximate; they're based on the Islamic lunar calendar and are subject to the first sighting of the moon.

Dairy Goats: Got Milk—or Cheese?

According to *Dairy Goats: Sustainable Production*, a livestock production guide issued by Appropriate Technology Transfer for Rural Areas (ATTRA), in 1994 there were 1 million dairy goats in the United States. These goats produced 600,000 tons of milk, which was marketed by three hundred dairy goat businesses and by at least thirty-five commercial goat cheese makers. Yet another 650 tons of goat cheese were imported from France alone.

Goat's milk and goat's milk products are in huge demand. If you have what it takes to succeed in the business, it's likely you can turn a profit milking dairy goats—but you have know what you're doing from the get-go.

Dairy businesses are labor intensive. It may be difficult to find reliable help at wages you can afford. If you can't, can you carry the operation by yourself? With your family's help? Is your family willing to help? Goats must be milked twice a day, at precisely the same hours, seven days a week—they never take holidays or weekends off. And milking is only part of the dairy worker's day. Goats must be fed and cleaned up after and doctored when sick or injured. Feed and bedding must be bought and stored. The milk room and milking equipment must be sanitized twice a day. One person shouldering the load can burn out fast.

Don't assume you can simply sell milk from your own back door. Most states enforce strict regulations governing the sale of fluid milk. Before setting up a business, be it milking ten goats or two hundred, you *must* contact the agency responsible for dairy regulation in your state and procure a license. Agencies vary from state to state. In

Organic Goat Meat

Health-conscious buyers are willing to pay a pretty penny for premium, "natural" meat. Organic chevon generally fetches the highest prices—but in most locales it's next to impossible to produce.

Why? In a word: parasites. Goat worms are legion, and goat producers must rely on chemical dewormers to keep them in line. But to market organic chevon, you must be enrolled in the USDA's Certified Organic Program, to which chemical dewormers are strictly verboten. Antibiotics and chemical-based medications (with a few exceptions) are as well. It's hard to raise goats organically, so thoroughly investigate certification before you commit—and research hardy breeds such as the Kiko, developed specifically for enhanced parasite resistance.

A better approach is marketing "natural" or "grass-fed" chevon. Such kids are raised without growth hormones or stimulants. They can consume browse, grass, and hay but not concentrates. Goats are naturals for this type of meat production.

Keep in mind, however, that unless you're licensed to sell processed meat, you must sell organic, natural, and grass-fed goats direct to your customers (alive) or delivered to a USDA-approved slaughterhouse.

This Saanen doe exhibits excellent dairy character (physical traits that suggest high milking ability). Dairy goats seem bony to the uninitiated but they're meant to be more angular than meat goats.

Arkansas, the Milk and Dairy Products division of the Department of Health is in charge; in Idaho, it's the Bureau of Dairying, part of the Idaho Department of Agriculture; in Colorado, it's the Consumer Protection division of the Department of Public Health and Environment. If you don't know where to look, ask your county extension agent or log on to the American Dairy Goat Association's Web site (see Resources), where you'll find up-to-date contact information for every state and several foreign countries.

You'll discover that in most states, to sell dairy products for human consumption (and especially raw milk), you'll have to set up a class A dairy. For this you will need: a milking parlor, a separate milk room, regulation equipment, and on-site waste system. The milking parlor must have a concrete floor (or one made of another impervious material); smooth, painted, or finished walls and dust-tight ceilings; approved lighting and ventilation; and metal or other nonwooden milking stands. The floor must slope away from the milk room.

There must be a separate milk room to house your bulk tank and cleanup area, and it must have a tight-fitting, self-closing door leading to the milking parlor. The milk room must incorporate the requisite ventilation, lighting, floors, and walls. A regulation hose port must be installed in one wall to transfer milk from your bulk tank to the milk transporter's tank, and your bulk tank must be installed according to strict agency specifications. A two-compartment wash sink with hot water under pressure is a must.

All milking equipment and the bulk tank must meet strict 3-A manufacturing standards. An approved on-site toilet and a dairy waste management system are also mandatory. For an in-depth look at typical equipment and operating requirements, visit "Grade A Dairy Goat Requirements" at Langston University's E (Kika) de la Garza Institute for Goat Research (see Resources).

Selling Milk

If you plan to sell fluid milk in bulk, investigate prospective buyers before you commit. In many locales, no such buyer exists. If you find one, contact a representative and ask if he or she needs additional suppliers. If so, how much milk will the buyer purchase from you? What is the payment? How much is charged for hauling? Are you expected to supply milk year-round? Request the names of existing suppliers, and contact all of them. Ask a lot of questions. Make sure the buyer is reliable before you buy goats (and expensive equipment!).

Before you sell milk or other dairy products direct from your farm, make absolutely certain it's legal. In most states, it's OK to sell to individuals who use the milk for animal food but not to individuals who use the milk for human consumption. Be careful. Fines for pedaling illegal milk are often very steep.

Another legal way to market your good goat's milk is through livestock.

Goat's milk, usually indistinguishable from cow's milk in appearance, is in high demand for drinking and for making cheese.

Pigs, calves, and meat goat kids thrive on goat's milk, and all are readily marketable as meat. Or market your milk as value-added products such as yummy, high-quality goat cheese.

Artisan Goat Cheese

In the United States, more than 250 specialty cheesemakers handcrafted millions of pounds of artisan cheese in 2002. Sales of specialty cheese topped $2.5 billion in 2000, up 4 percent from 1999, and they're expected to rise an additional 4 percent per annum through 2005. The best part: much of that cheese is made all or in part from goat's milk.

When most people think of goat cheese, they visualize *chevre*, the tasty, tart, earthy goat cheese from France.

How Much Milk Do Milk Goats Give?

Does in Milk for 275–305 Days	Number of Does	Milk (lb/average)	Range (lb)	Butterfat % lb
Alpine	699	2,254	840–5,300	3.5 78
LaMancha	216	2,097	1,050–3,510	3.9 81
Nubian	445	1,746	640–3,670	4.8 84
Oberhasli	68	2,062	990–3,629	3.7 76
Saanen	432	2,468	970–5,630	3.4 84
Toggenburg	184	2,015	860–4,480	3.2 64

ADGA Averages for 2002 Lactations
Adapted from *Dairy Goats: Sustainable Production* (ATTRA)

But goat cheese runs the gamut from spoonable, silky *fromage blanc* to mold-coated, Brielike *chevrita*, to cheddarlike firm, ripened cheeses.

You won't learn to make quality artisan goat cheeses overnight; you need to apprentice to an established cheesemaker, attend seminars, or take courses. By doing so, you can learn to produce an imminently salable, in-demand product marketable at farmer's markets, to restaurants, through retailers, or from home via mail-order or online sales.

Before undertaking any goat-related enterprise, contact ATTRA (Appropriate Technology Transfer for Rural Areas; see Resources). Although you can read and download many of their bulletins and handbooks online, if you call and speak with an ATTRA adviser, he or she will compile a free information packet tailored specifically for your needs. Don't overlook this resource! If you'd rather peruse ATTRA handbooks online, check out *Goats; Sustainable Production Overview*, *Dairy Goats: Sustainable Production*, and *Value-Added Dairy Options* among ATTRA's many value-added marketing, pasture management, and organic production bulletins and booklets.

You'll also want to download the *Small Dairy Resource Book* from the Sustainable Agriculture Research and Education (SARE) Web site (see Resources). Simply click on *Publications* and scroll down to the title. It covers essential topics on processing and marketing of dairy products.

In addition, visit the Maryland Small Ruminant Page's Web site, where dozens of links lead you to resources as

Household Dairy Goats—Are They for You?

If you don't want to sell dairy products but you'd like to stop buying them at the store, get a goat! A few goats can keep a four-person family in milk, cheese, and yogurt year-round. Dairy goats are relatively inexpensive and simple to milk and to maintain.

Though dairy breeds were developed for milk production, any type of goat—even meat and fiber does—will give delicious milk, just not as much. Many owners swear by the rich, creamy milk produced by Pygmy and Nigerian Dwarf goats. What the little girls lack in quantity they make up for in lip-smacking goodness. Whatever the type, buy your goat from a responsible, knowledgeable breeder or dairy. Ask for copies of her production records, and insist on seeing her milked before you buy. In fact, ask the seller to show you how to milk the doe. Although you can buy milking machines set up for one or two goats (costing $1,300–$1,400 from most dairy equipment suppliers), hand milking a gentle doe can be very relaxing. If you haven't milked before, start with trained goats. Goats are smart and wily, so you'll want to avoid dealing with a doe who doesn't want to be milked.

Milk must be handled properly to eliminate off-flavors. You'll spend time quick cooling, filtering, and probably pasteurizing your goat's milk; to do it, you'll need the right equipment. Send for free catalogs from Hoegger Goat Supply and Caprine Supply (see Resources) to see what milking entails. It's not difficult, but it does take some time and effort. Is it worth it? We think so, but educate yourself, then you can decide.

Two young Norwegian girls milk the family goat in bygone days. Dairy farmers wanting to sell goat's milk for human consumption today must forgo this casual approach and brisk outdoor setting for a strictly regulated milking parlor.

diverse as the French CIRVAL resource center for dairy sheep and goats, University of Vermont's Small Ruminant Dairy Project (subscribe to their free newsletter), the American Cheese Society, and the resource-rich small dairy.com cheese makers network.

Finally, consider subscribing to YahooGroups dairy goat- and small dairying-interest lists to learn firsthand from hundreds of folks already involved in the dairy goat business (see Resources). After digesting these resources and doing the math, you'll have a clear idea of what it takes to succeed in the goat dairy biz and if it's right for you.

Fiber Goats: The Chosen Ones

Two types of fiber are harvested from goats: mohair and cashmere. Angora goats produce mohair, as do some pint-size Pygoras, and almost any breed or type can yield cashmere. Let's take a peek at both kinds.

Belying their names, Angora goats produce mohair, not angora (check with your local rabbits for the latter).

Mohair

It stands to reason that Angora fiber would be clipped from Angora goats—but it isn't. Angora is the hair of Angora rabbits. Angora goats produce *mohair*, a word derived from the Arabic word *mukhaya*, meaning "the chosen." Sumerian tablets and biblical references set the breed's development sometime between the fifteenth and twelfth centuries BC. Its name, Angora, is derived from Ankara, a Turkish city where the Middle Eastern mohair industry evolved. Mohair became so important to the Turkish economy that none of its precious goats was sold abroad until the sixteenth century AD, when Charles V of Spain brought the first pair of Angoras to Europe. America's first Angoras arrived in 1849, a gift from the sultan of Turkey to a Dr. James B. Davis of South Carolina. Since then, Angora numbers have skyrocketed.

Angoras' manageable size and calm dispositions make them easy to handle. Adult does weigh 70–110 pounds; bucks average 120–185 pounds. Both sexes grow slowly, rarely maturing until they're three to five years old. Angoras breed seasonally. They produce longer than other breeds do; nine to fourteen years is not unusual (well cared for Angora goats can live to eighteen years of age). Compared with other breeds, lower birthing rates are the norm. According to the Mohair Council of America, 60–70 percent is average for large commercial herds. In well-managed hobby farm settings, 100–120 percent is possible. Most does who carry a pregnancy to term produce a single kid, although twins aren't uncommon.

According to the National Agriculture Statistics Service, as of January 1, 2005, there were 274,000 Angora goats in the United States, most located in Texas (210,000, in fact; 90 percent of these located within a 150-mile radius of San Angelo). Texas is the long-time heart of the American mohair industry and home of the Mohair Council of America. Furthermore, in 2005, some 269,500 Angora goats were shorn, producing an average per clip of 7.2 pounds. Commercial-class mohair averaged $1.97 a pound, for a total of 3.8 million dollars.

However, conscientious hobby farmers can earn considerably higher income by producing Angora goat fleeces for the burgeoning handspinner's market. Handspinning is taking the country by storm. Handspinners are paying $3–$15 a pound for white mohair fleece, and $6–$20 per pound for the fleece of colored Angora goats. These prices are for unwashed but judiciously trimmed, clean Angora goat fleeces. Since mohair becomes coarser as goats age, highest prices are for kid fleeces (generally $12–$20 per pound). The trick to greater profits is keeping handspinner-quality fleeces free of muck, manure, dirt, and bits of organic matter such as hay chaff, twigs, and burrs; it's almost impossible in a huge flock setting, but manageable for the hobby farmer with only thirty goats. Fleece can be damaged by external parasites, especially lice. Like most breeds, Angoras are highly susceptible to internal parasites.

Angoras come in pure white and colored versions. Colored Angoras are currently the rage. The big-industry fiber buying co-ops won't buy colored fiber,

Why Mohair? Why Cashmere?

- The chemical composition of mohair and cashmere is similar to that of wool, but they have much smoother surfaces, so they lack the felting properties of wool.
- White cashmere and white mohair accept dye exceptionally well.
- Soft, strong, lustrous, and elastic mohair yarn is woven into garments of all sorts. The world's finest teddy bears and similar stuffed animal toys are crafted of mohair fabric, whereas dolls' hair and the like are often made of natural, flowing locks of mohair.

The abundant curly white hair of this photogenic goat, who patiently poses with tiny driver and passenger, leaves no doubt that you're viewing an Angora goat. Using an Angora as a harness animal probably isn't a good idea if you want to sell that fleece for a good price.

but handspinners prefer it. Colored breeding stock costs more than white does, but all things equal, colored fiber sells at a higher price per pound. Angoras must be fed a nutritious diet to produce quality fleece. Angoras are excellent browsers and brushers, but they definitely require supplemental feeding.

Angoras aren't as winter hardy as other breeds are; they do best in semi-arid regions such as Texas. Angoras can be raised as far north as Minnesota, Wisconsin, Upper Michigan, and New England, but suitable winter shelters are an absolute must, especially at kidding time and after shearing.

These goats *must* be clipped, preferably twice a year. Scissors and manual sheep shears can be used to clip small flocks, but larger operators use electric sheep shears (not horse- and cattle-style clipper-head units) fitted with a twenty-tooth goat comb. They are shorn twice a year, generally in March (prior to kidding) and again in the fall, yielding 5–10 pounds of 4–6 inch, wavy locks per clip. A well-managed Angora annually yields 20–25 percent of its body weight in mohair, making it the most efficient fiber animal in the world. They're also popular as pets and as 4-H/FFA project goats.

Pygora Fiber

Pygoras are cute, scaled-down fiber goats developed by Oregon goat breeder Katherine Johnson, who bred Pygmy goats to full-size Angoras. Pygora does weigh 65–75 pounds and stand at least 19 inches tall; bucks and wethers tip the scale at 75–95 pounds and are 23 inches or taller. They come in all Pygmy goat colors and their dilutions, plus white. Each Pygora produces fleece of one of three types:

- Type A (Angora type)—lustrous, curly fiber up to 6 inches long. These animals must be shorn.
- Type B (blend type)—a blend of Pygmy goat undercoat (cashmere) and 3–6 inches Angora mohair. These goats shed and can be shorn, plucked, or combed.
- Type C (cashmere type)—very fine 1–3 inches fiber with no luster. These goats shed and may be shorn or combed.

Cashmere

Cashmere goats are not a breed. Cashmere is the winter undercoat (down) produced by nearly every breed of goat. Cashmere goats are simply goats who produce a bountiful supply of undercoat, especially undercoat in the acceptable micron range.

Fiber diameter is measured in microns; a micron is one-millionth of a meter. Mohair measures 20–50 microns; Merino sheep wool, the finest wool of all, 18–25 microns; alpaca, 15–35 microns; and cashmere, just 12–19 microns, making cashmere one of the cushiest animal fibers on earth.

Undercoat growth begins about midsummer and stops around the winter solstice. Cashmere-producing goats can be sheared prior to their natural shedding time (the undercoat is separated from guard hair by a commercial dehairing machine) or combed as the animals shed. A productive goat yields 4 ounces of guard hair–free cashmere fiber. Clean cashmere fiber more than 1.25 inches in length and of finest micron count currently fetches about $30 per pound; shorter cashmere, roughly $7–$10 per pound.

Most cashmere fiber is produced in China, 3,000 tons annually; Mongolia produces another 2,000 tons. America's cashmere industry is emerging as breeders realize that Spanish goats, dairy and meat breeds, and even some Boers produce cashmere worth shearing, making cashmere a money-making secondary cash crop. About 150 American producers are currently harvesting two thousand head of goats. It's a start!

Did You Know?

Because of their cushy softness, elegant drape, and hefty price tags, cashmere garments have always appealed to affluent buyers. Louis Bonaparte presented the first Indian pashmina cashmere shawl in France to Madame Bourienne in 1799. He started a trend. Emperor Napoleon Bonaparte later gifted Empress Eugenie with a total of seventeen pashmina shawls. Queen Victoria loved these shawls, too; her wearing them made pashmina shawls popular throughout Great Britain. Early nineteenth-century English dandy Beau Brummel was famous for his white cashmere waistcoats. The "original sweater girl," Lana Turner, wore a tight cashmere sweater in a 1937 film called *They Won't Forget*; due to her influence, cashmere sweaters remained popular through the 1950s.

Whether you own only a few goats or a large herd, goat keeping can be both a profitable and a rewarding venture.

Acknowledgments

Thanks again to the good folks who contributed "Advice from the Farm" tips and words of wisdom and photographs.

Jerry and Lyn Brown raise registered California Red Sheep on their Shear Perfection Ranch in the La Plata Valley of the Four Corners area of New Mexico. Visit the Shear Perfection Web site at http://www.nmredsheep.meridian1.net, and contact Lyn at info@nmredsheep.meridian1.net.

Carl Langle is one of only fourteen American breeders of full-blood Savanna goats. Contact the Langles at 1658 Liberty Street, Viola, AR 72583, 870-458-2140.

Mona Enderli and her husband, Joey, live near Baytown, Texas, where they breed full-blood and percentage Boer show wethers. They created and sponsor the "Extreme Show Goat Team" made up of kids from surrounding counties who compete on the market goat show circuit. Both are AMGA-certified meat goat judges. Contact the Enderlis at 281-421-8073 or enderlifarms@hotmail.com, and visit their Web site at http://www.enderlifarms.net.

Donna Haas is a nice lady from Missouri who loves all animals, especially her home-raised part-Boer goats.

Rikke Giles and her husband raise purebred Nigerian Dwarf goats, farm fresh vegetables, herbs, and flowers on FoxDog Farm in Kingston, Washington. Visit their Web site at http://www.foxdogfarm.com, and contact Rikke at rgiles@centurytel.net.

Matt Gurn and his wife, Claudia Marcus-Gurn, live near Winona, Missouri. They raise show-quality Boer goats and easy-gaited horses. Visit their information-packed MAC Goats Web site at http://members.psyber.com/macgoats, and contact them at 417-778-1904 or at macgoats@ortrackm.missouri.org.

Melody Hale raises Nubian goats, Jacob sheep, and assorted poultry at CritterLand Farm in central Oregon. She is interested in the preservation of endangered breeds of domestic livestock and heritage poultry, and she is a member of the American Livestock Breeds Conservancy. Contact her at critterland@bendcable.com.

Samantha Kennedy lives on the 4TS Ranch in Delta, Colorado, with her husband, Todd, and their three sons, Ty, Tanner, and Tegan. The Kennedys raise full-blood, purebred, and percentage Boer goats and a few special LaManchas. They sell breeding stock, butcher stock, and show wethers. Visit their Web site at http://www.webspawner.com/users/4tsboergoats/index.html, and contact Samantha at 970-874-8056.

Six years ago, while *Bobbie Milsom* was recovering from intricate surgery required to mend her broken neck, Bobbie's daughter presented her with her first Pygmy goat. Nowadays she raises six breeds of goats and two breeds of sheep on her ranch near Maricopa, Arizona. Visit her Arizona Pygmy Goats Web site at http://www.arizonapygmygoats.com and her Arizona's Little Cudchewers site at http://littlecudchewers.tripod.com, or send e-mail to Bobbie at cudchewers@peoplepc.com.

Glenda Plog lives in Queensland, Australia. She has owned Angora, Cashmere, and Anglo Nubian breeds and now works with meat goats because they're less work. She has an applied science degree in rural technology and is working on a master's in animal studies, majoring in caprine helminths and the selection of parasite resistant animals for breeding stock. Contact Glenda at Glenjoy@uq.net.au.

Alyssa Rockers lives near Carthage, Missouri. She raises and shows full-blood and percentage Boer goats. Contact her at Rocky Acres Boer Goats 417-358-1778 or drockers@joplin.com.

Dave and Dixie Rockers raise full-blood and percentage Boer goats near Carthage, Missouri. Contact the Rockers at Rocky Acres Boer Goats 417- 358-1778 or drockers@joplin.com.

Lisa Shumack lives in northeastern Pennsylvania. She and her family raise and show Sable and Saanen dairy goats and use their excess milk to raise pigs and a calf for home consumption. Lisa is currently learning to craft goat cheese and goat milk soaps.

Pat Smith has had Alpine dairy goats since 1965; she joined the ADGA that same year. Pat lives in Hebron, Indiana. Contact her at andvell@webtv.net.

Kari Trampas lives near Seymour, Missouri, where as Christie's Dairy Goats, she and her daughter milk, raise, and show Saanens, Sables, LaManchas, and Nubians. As Cedar Hills Farm Boers, they breed percentage Boer goats, including paints, blacks, and reds. Visit their Web site at http://karitrampas.tripod.com or contact Kari at 417-935-2553 or saanens@fidnet.com.

Robin Walters lives near Seguin, Texas, where she raises Boer meat goats and operates Bar None Web Site Design and Management—no wonder her Bar None Meat Goats site (http://www.barnonemeatgoats.com) is one of the best online! Contact Robin at 830-401-5867 or barnone@gvec.net.

Michelle Wilfong lives in western Pennsylvania. She raises Nubian dairy goats and Myotonic (Fainting) goats. You can visit Michelle's Web site at http://www.Griffin HillFarm.com, or call her at 724-513-398.

Appendix: Goat Diseases at a Glance

Abortion

• Enzootic abortion (EAE) of does is a chlamydial disease transmitted from aborting goats and fetal tissues to other does. Infected does abort during the last month of pregnancy or give birth to stillborn or weak kids who soon die. An effective vaccine is available. (The second *E* in the abbreviation stands for "ewes," but this is a problem for goats as well.)

• Vibrosis is caused by the bacterium *Campylobacter fetus*, subspecies *intestinalis*. When one or two does affected by vibrosis abort, they can trigger an "abortion storm." Vibriosis vaccine is available, often in combination with EAE vaccine.

• Toxoplasmosis, which is caused by the coccidium *Toxoplasma gondii*, is spread when a host cat contaminates goat feed and water with her droppings. There is no vaccination or treatment for toxoplasmosis.

• When a doe aborts her kid, the fetus and tissues should be submitted to a laboratory for diagnosis; you can't treat the rest of the herd unless you positively know what's wrong. Your vet can tell you where to send the specimens. The material must be fresh, so store it in sturdy plastic bags, pack the bags in a Styrofoam box and surround them with chill packs, then rush the package to the lab.

Bloat

• Bloat is a buildup of frothy gas in the rumen.

• Bloat is usually triggered when a goat tanks up on an unaccustomed abundance of grain, rich grass, or legume hay.

• Bloated goats can quickly die of the condition, so if you suspect that your goat has bloat, call your vet posthaste.

Caprine Arthritis Encephalitis (CAE)

• Caprine Arthritis Encephalitis (CAE) is an incurable viral infection caused by a retrovirus similar to the one that causes HIV in humans. CAE infects only goats.

• A relatively uncommon juvenile-onset, neurological form of CAE causes encephalitic seizures and paralysis in kids, but CAE is primarily a wasting disease of adult goats. Early symptoms of infection include swollen knees, unexplained weight loss, and congested lungs. Sufferers eventually die of chronic progressive pneumonia.

• Initially associated mostly with dairy goats, CAE is spreading due to the widespread practice of crossing meat breed bucks with dairy and part dairy percentage does to produce commercial meat goats.

• Since CAE is spread from infected does to her kids via body fluids, colostrum, and milk, producers are breaking the cycle by removing kids from their dams at the moment of birth and then artificially rearing them either on pasteurized milk or on milk replacer.

• CAE testing of individual goats is possible, but as these tests aren't 100 percent accurate, it's best to buy from certified CAE-free herds.

Caseous Lymphadenitis (CLA)

• Caseous lymphadenitis (CLA) is a chronic, contagious disease of sheep and goats caused by the bacterium *Corynebacterium pseudotuberculosis*. The bacterium breaches a goat's body through mucous membranes or via cuts and abrasions. The animal's immune system valiantly tries to localize the infection by surrounding it in one or more cysts. If the ploy is unsuccessful, he will die.

• CLA presents as lumps near the jaw, in front of the shoulder, and where a doe's udder attaches to her body. Some goats develop internal cysts, too.

Coccidiosis (Cocci)

• Coccidiosis is a very common, potentially fatal yet easily prevented, easily treated disease of young kids. It's caused by uncontrolled proliferation of single-cell protozoal parasites called coccidia, found in barnyard soil. Cocci is species-specific, meaning goats aren't bothered by poultry, canine, or sheep coccidia.

• Suspect cocci when kids more than two weeks old experience severe abdominal pain (evinced by crying or reluctance to lie down) coupled with dark, watery, foul-smelling diarrhea streaked with mucus or blood. Take a stool sample to your vet for fecal diagnosis.

• Although rehydration with electrolyte solutions and administration of antidiarrheal medications along with sulfa drugs, amprolium, or tetracycline usually effect a cure, it's easier to prevent cocci than it is to cure it. Many producers choose to give their goats feeds laced with anticoccidial drugs called coccidiostats, whereas others add them to the goat's drinking water. Horse owners, take note: one such feed additive, Rumensin (monensin), is extremely toxic to horses.

Contagious Ecthyma (CE)

• Commonly known as sore mouth, also known as scabby mouth or orf, contagious ecthyma (CE) is a contagious poxlike virus that causes the formation of blisters and pustules on the lips and inside the mouths of young kids, as well as on the teats of the infected kids' mothers. The blisters pop, causing scabbing and pain so intense that occasionally a kid will starve rather than eat. Most kids recover in one to three weeks without treatment.

• Because sore mouth is easily transmissible to humans, wear rubber gloves when handling stricken kids. Keep children away from all infected goats!

• An effective live vaccine is available, but you mustn't use it unless you already have sore mouth on your property. Goats will shed their vaccinations scabs, which will contaminate your property and almost certainly spread the disease to the rest of your herd.

Enterotoxemia

• There are two types of enterotoxemia in goats caused by *Clostridium perfringens*: Types C and D.

• Type C is a disease of young kids caused by an anaerobic bacterium found in manure and soil. It enters via newborn kids' mouths when they encounter dirty conditions while seeking their mothers' udders. Bacteria produce a toxin that causes rapid death. Treatment is usually ineffective; death usually occurs within two hours of the onset of symptoms, which include seizures and frothing at the mouth. However, kids from does vaccinated for enterotoxemia during late pregnancy develop immunity to the disease via their mothers' colostrum.

• Type D is also present in the soil and manure. It attacks rapidly growing, slightly older kids who ingest the bacterium while investigating their environment. It, too, causes rapid death and with it tremors, convulsions, and a host of strange neurological behaviors. A vaccine is available alone or in combination with type C or as a C/D and tetanus vaccine.

Floppy Kid Syndrome (FKS)

• Floppy kid syndrome (FKS) is a relatively new malady that affects kids between three and ten days old. Its precise cause is still uncertain.

• Kids autopsied as part of a study conducted by Texas A&M University had very distended abomasums full of acidic-smelling, coagulated milk. Scientists speculate that overconsumption of rich milk triggers an overgrowth of certain microorganisms in the digestive tract, resulting in systemic, often deadly, acidosis.

• Afflicted kids show muscular weakness and depression, progressing to flaccid paralysis, and often death. In all cases, their abdomens are distended, and if gently shaken, they may "slosh."

Johne's Disease

• Johne's (YO-neez) is a deadly, contagious, slow-developing, antibiotic-resistant disease affecting the intestinal tracts of domestic and wild ruminants, including goats.

• The bacterium that causes Johne's, *Mycobacterium avium,* subsp. *paratuberculosis,* is closely related to the one that causes tuberculosis in humans. Infected goats are dull, depressed, and thin. Johne's disease, also known as paratuberculosis, is incurable.

Ketosis

• Ketosis is a relatively common metabolic condition associated with pre- and postpartum does, especially overweight and underexercised does pregnant with more than one kid.

• Prepartum ketosis is also called pregnancy toxemia; it occurs within a month before birth. The rumen of a fat doe carrying multiple kids is scarcely able to hold enough nourishment to meet the nutritional needs of her late-term kids, so her body begins burning her own fat reserves to provide energy.

• Postpartum ketosis, also called lactational ketosis, occurs when the high energy demand on a doe nursing multiples (especially triplets, quads, and quints) causes excessive weight loss—she simply can't consume enough feed to meet their needs, so she dips into her own reserves.

• In both scenarios, ketones produced by this process make the doe quite ill. Without intervention she'll die. Early symptoms include listlessness, poor appetite, and possibly labored breathing, progressing to circling, stargazing, stumbling, and teeth grinding. She'll eventually collapse and, without aggressive treatment, lapse into a coma and die.

• To prevent life-threatening ketosis, does should be fed a high-quality, balanced diet throughout their pregnancies and monitored closely the month before and the month after giving birth.

Listeriosis and Goat Polio

• Listeriosis (also known as circling disease) and goat polio (also called polioencephalomalacia and cerebrocortical necrosis) are serious metabolic diseases with similar causes and similar symptoms. Both of these diseases occur mainly among confinement-kept goats who are fed relatively high-concentrate, low-fiber diets, and both of these diseases can be triggered by abrupt changes in feed and by moldy grain or moldy forage (especially silage).

• Listeriosis is caused by the common bacterium *Listeria monocytogenes*. One type of listeriosis causes abortions; the other type, which is more common, causes encephalitis. Both types are usually seen in adult goats. Encephalitic listeriosis triggers inflammation of the brain stem and death (necrosis) of brain tissue, resulting in one-sided facial paralysis, drooling, stargazing, and stumbling; lack of appetite, depression, and fever are other common symptoms. Listeriosis can be passed to humans via the milk of sick or carrier goats. Without aggressive antibiotic treatment, afflicted goats die.

• Goat polio is a deficiency of vitamin B1/thiamine, most commonly encountered in weanling and yearling goats. Ingesting excess amounts of grain, stress, prolonged or excessive use of antibiotics, or changes in feed drastically lower rumen pH, and beneficial microorganisms die off. This, in turn, decreases thiamine production. Thiamine is necessary to metabolize glucose. Without glucose to feed them, brain cells die and neurological symptoms such as hyperexcitability, staggering or weaving, circling, blindness, and tremors appear. Un-treated goats develop convulsions and usually die in one to three days; promptly injected with thiamine (best given intravenously by a vet), goats usually recover.

Mastitis

• Mastitis is a serious infection of the mammary system. Substandard milking hygiene, delayed milking, and injuries are common causes. Symptoms include decreased milk production; clumps, strings, or blood in milk; and pain, inflammation, and swelling in the udder. Usually only one side is affected.

• Does must be milked in sanitary surroundings, and hands and udders must be thoroughly cleansed before and after milking. Milk gently. Milk at scheduled intervals. Use home-based mastitis test kits to check for mastitis at weekly intervals.

• If mastitis is suspected, seek professional treatment. Untreated mastitis swiftly leads to permanent udder damage—one type leads to gangrene and death.

Pneumonia

• Pneumonia is caused when one of a wide variety of opportunistic bacteria and viruses mix with stressed goats.

• Typical symptoms include depression, fever, coughing, and labored breathing. Because so many bacteria and viruses may be involved, accurate identification of the infectious agent is an essential part of successful treatment.

Scrapie

• Scrapie is a transmissible spongiform encephalopathy (TSE) of sheep and goats similar to bovine spongiform encephalopathy (BSE, or mad cow disease) and to chronic wasting disease (CWD, which affects deer and elk). No human has ever contracted scrapie (or either of the human equivalents, kuru and Creutzfeldt-Jakob disease) from sheep or goats.

• Goats residing on properties where sheep are also present must be identified through the USDA's mandatory scrapie eradication program. At press time, provisions were rapidly changing, so contact your state Animal and Plant Health Inspection Service (APHIS) representative for up-to-date information.

• Scrapie appears to be caused by an infectious agent, but genetics also play a part. The disease was recognized in Britain and western Europe at least two hundred years ago, and it came to the United States in 1947 with British goats. Scrapie is a global scourge: only Australia and New Zealand are scrapie free.

• Scrapie is a slow, progressive disease that systematically destroys the central nervous system. It is far more prevalent in sheep than it is in goats. Symptoms typically appear two to five years after contraction and include weight loss, hypersensitivity, tremors, stumbling, blindness, excess salivation, lip smacking, and intense itchiness. Between one and six months after symptoms appear, infected animals die.

Urinary Calculi

• Urinary calculi are tiny stones or crystals that form in the urinary tracts of sheep and goats. Does get stones, but they pass through the larger female urethra (the tube that empties urine from the bladder) without difficulty. A buck or wether with a blocked urethra is in trouble, however; his bladder is apt to rupture, and he'll probably die.

• When bucklings are castrated, penis growth stops, so wethers castrated at an early age are especially troubled by calculi; their much tinier penises and urethras are easily blocked. A workable solution: don't castrate male kids younger than four to six weeks old.

• A calcium-phosphorus ratio of 2:1 in the diet helps prevent calculi formation, as do small, measured amounts of ammonium chloride added to feed. Male goats should drink lots of water. Make it more appealing by keeping water sources readily available, full, and sparkling clean.

White Muscle Disease

• White muscle disease is caused by selenium deficiency.

• Does grazing on selenium-poor land or those eating hay that was raised in such depleted conditions require selenium/vitamin D supplementation during the last two months of pregnancy. Otherwise, their affected kids will have problems rising and walking. Some kids will even become paralyzed. Prevention is the key to eliminating white muscle disease.

Glossary

Abomasum—the fourth compartment of the ruminant stomach

Accredited herd—one annually tested for tuberculosis and certified TB-free

Afterbirth—fetal membranes expelled after the birth of kids

AI—artificial insemination

American—in American Dairy Goat Association terminology, a fifteen-sixteenths purebred buck or a seven-eighths purebred doe (e.g., American Saanen)

Artificial rearing—raising kids on milk or milk replacer

Band—(noun) a strong rubber band used to castrate kids; (verb) the act of using an elastrator to apply one of these bands

Billy—an outdated word denoting an uncastrated male goat; today's goat fanciers and breeders strongly discourage the use of this term

Blind teat—a nonfunctional teat

Body condition score—a rating from 1 (very thin) to 5 (obese) used to estimate the condition of goats

Bolus—a large, oval pill

Breech birth or breech delivery—one in which a kid's hind feet come first

Broken mouth—an old goat with missing or broken teeth

Browse—(noun) edible woody plants such as twigs or saplings and wild berry canes; (verb) the act of eating browse

Buck—an uncastrated male goat

Buckling—an uncastrated male kid

Buck rag—a cloth that is rubbed on a buck's scent glands, then kept in a closed jar to hold the scent, and later presented to a doe to determine if she is in heat

Burdizzo—a tool used to castrate bucks and bucklings by severing the cord without breaking the skin of the scrotum

Butt—(verb) to bash another goat, a human, or an object with the forehead or horns

Cabrito—(Spanish, "little goat") the meat of pink-fleshed, milk-fed kids

Capretto—(Italian, "little goat") same as cabrito; sometimes used in international goat recipes

California mastitis test (CMT)—an easy, do-it-yourself home mastitis test

Caprine—having to do with goats

Caprine Arthritis Encephalitis (CAE)—see the Appendix

Caseous Lymphadenitis (CL)—see the Appendix

cc (cubic centimeter)—a unit of fluid medication measure equal to one milliliter

Certified herd—one annually tested for brucellosis and certified free of specific serious diseases

Chevon—(French, "goat meat") any type of goat meat, especially that of fairly mature goats

Chevre—soft French cheese crafted of goat's milk

Club kid—a kid raised as part of a 4-H or an FFA project

Coccidiosis (Cocci)—see the Appendix

Colostrum—a doe's first milk; it contains antibodies that protect her kids through their first few months of life, at which point they develop disease resistance of their own

Concentrates—the nonforage portion of a goat's diet; particularly grains, meals, and commercial goat feed

Confinement housing—the act of confining goats to a barn and exercise area in lieu of keeping them on pasture

Conformation—an animal's physical characteristics

Creep—a feeder designed to allow kids to enter and eat while keeping larger goats out

Crossbred—an animal with parents of two different breeds

Cud—a glob of regurgitated food that's rechewed and swallowed again

Cull—(verb) the act of removing undesirable goats from a herd; (noun) a goat removed as part of the culling process

Dairy Herd Improvement Association (DHIA)—a milk testing program administered at state levels, under the jurisdiction of the USDA

Dairy Herd Improvement Registry (DHIR)—a milk testing program administered by dairy goat registries in cooperation with the USDA

Dam—an animal's female parent

Dehorning—the grisly removal of horns from an adult goat

Dental palate (dental pad)—a goat's firm upper palate

Disbud—to destroy a very young kid's horn buttons by burning them with a hot iron

Disbudding iron—the electric or fire-heated tool used to disbud young kids

Dish face—a concave profile common in Pygmy goats and Swiss dairy breeds

Doe—a female goat

Doeling—a female kid

Drench—(noun) liquid medicine given orally; (verb) to administer a drench

Dry doe—a doe between lactations

Dry off—to cease milking a doe

Dual-purpose breeds—breeds developed for both milk and meat production (Nubians are dual-purpose goats)

Elastrator—a tool used to apply thick rubber bands to bucks' or bucklings' scrotums for castration

Elf ear—a type of LaMancha goat ear that is up to 2 inches long

Emasculator—a tool used for docking and castrating bucks and bucklings; it has a crushing effect, which helps reduce bleeding

Ennobled—an honors classification attainable by registered Boer goats

Enterotoxemia—see the Appendix

Estrus—heat; the period during which a doe is receptive and can conceive

Fleece—raw fiber, usually in one piece, as shorn from a single fiber goat

Flushing—increasing a doe's nutritional level prior to breeding season

Fly-strike—a condition caused when blowflies lay eggs in wounds or wet, filthy fleece; maggots develop and consume the host's flesh

Footbath—a chemical mixture that goats walk through or stand in, designed to prevent or treat hoof disease

Forage—fibrous animal feeds such as browse, grass, and hay

Free choice—method of feeding when food is made available 24-7

Freshen—to give birth and come into milk

Gestation—the period of pregnancy beginning at conception and ending at birth

Goatling—(British) an older kid

Gopher ear—a type of LaMancha goat ear that is free of cartilage and is 1 inch or less in length

Graft—the act of persuading a doe to adopt another doe's kid or kids

Guard dog (llama, donkey)—an animal who bonds with and stays with goats to guard them from predators such as coyotes, wolves, bears, cougars, and eagles

Gummer—an old goat with no teeth

Halter—headgear used to lead or tie an animal

Handle—how fleece feels to a spinner

Heat—estrus; the period when a doe is receptive to a buck and can conceive

Hermaphrodite—a sterile goat having both male and female reproductive organs

Hog butt—the heavily muscled hindquarters of well-conformed meat breed bucks

Horn buds—two small forehead lumps from which kids' horns emerge

Hybrid vigor—the extra strength, hardiness, and productivity exhibited by animals whose parents are of two different breeds

Intramuscular injection (IM)—an injection inserted into muscle

Intravenous injection (IV)—an injection inserted into a vein

In kid—pregnant with kids

In milk—lactating, giving milk

Johne's Disease—see the Appendix

Jug—a cozy mothering pen used by a single doe to bond with and watch over her newborn kids in peace

Keds—bloodsucking, wingless flies sometimes called sheep ticks

Ketones—compounds found in the blood of pregnant does suffering from pregnancy disease (ketosis)
Kid—a baby goat of either sex
Kidding—giving birth to kids
Lactation—the period during which a doe produces milk
Let down—release of milk by the mammary glands prior to milking
Linebreeding—the breeding of closely related goats; used to fix the type and intensify the characteristics of shared ancestors
Liver flukes—tiny leaf-shaped parasites that dwell in bile ducts and liver tissue
Livestock guardian dog (LGD)—a dog of specific livestock guardian breed background that lives with goats and protects them from predation
Lungworms—parasites that infest the respiratory tract and lung tissue
Mastitis—serious inflammation of the udder
Mating capacity—the number of does a buck can impregnate in a season
Meconium—the first manure passed by a newborn kid
Metritis—inflammation of the uterus
Microorganisms—microscopic creatures; bacteria, protozoa, and the like
Milking stand—an elevated platform fitted with a head stanchion, upon which a doe or dairy ewe stands to be milked
Milking through—milking a doe for more than one year
ml (milliliter)—a unit of fluid medication measure equal to one cc
Nanny—an outdated word denoting a female goat; today's goat fanciers and breeders strongly discourage the use of this term
Nose bots—larvae of the botfly living in the nasal passages of a goat
Omasum—the third compartment of the ruminant stomach
On test—enrolled in the DHIA milk testing program
Open doe—a doe who isn't pregnant
Orifice—the opening to a teat
Ovulation—the period when an egg is released from an ovary and a doe can conceive
Oxytocin—the hormone that controls milk letdown; oxytocin shots are sometimes given to help does expel afterbirth tissue
Papered—registered
Papers—registration certificates
Pasture kidding—the act of allowing does to give birth at pasture instead of in a barn
Pedigree—an animal's "family tree"
Pelt—a hair-on, tanned goat hide
Percentage—crossbred; a term used among breeders of meat goats to denote how much Boer, Kiko, or other breed appears in a crossbred's pedigree
Placenta—afterbirth; fetal tissue expelled after a kid is delivered
Polled—naturally hornless
Precocious milker—a doe who produces milk without being bred
Probiotics—oral gels and powders (such as Probios and Fast Track) fed to goats to help repopulate their rumens with beneficial bacteria that were lost due to antibiotic therapy, disease, or stress
Progeny—offspring

Purebred—an animal whose ancestors for a set number of generations were registered and all of the same breed

Recorded goat—a partbred or crossbred dairy goat whose pedigree and particulars are recorded in a dairy goat registry's official herdbook

Registered goat—a purebred goat whose pedigree and particulars are registered in a registry's official herdbook

Rennet—an enzyme used to set curds for cheese making

Reticulum—the second segment of the ruminant stomach

Roman nose—the arched profile typical of Nubian, Boer, and Savanna goats

Rumen—the large first stomach compartment of a ruminant where feed is broken down into usable elements

Ruminant—a cud-chewing animal with a four-compartment stomach

Ruminate—the act of chewing cud

S/T/Tr—shorthand for single, twin, and triplet births

Scours—diarrhea

Scrapie—a serious, transmissible spongiform encephalopathy malady much like mad cow disease

Scrub goat—goat of mixed ancestry commonly used for land clearing; also called a brush goat

Scur—a misshapen horn caused by improper or failed disbudding

Septicemia—an infection of the bloodstream that affects the entire body

Settle—to get pregnant

Shearing—the act of removing fiber from Angora and Pygora goats

Silent heat—being in heat without showing outward signs

Sire—an animal's male parent

Stanchion—a head restraint used to contain does while milking them

Standing heat—the period during which does are receptive to the buck

Stocking rate—the number of animals grazed on an acre of land

Subcutaneous injection (SQ or SubQ)—an injection inserted directly under the skin

Tag—(noun) a dreadlock of manure-laden fiber; (verb) the act of clipping tags from a fiber goat's fleece

Tapeworms—long, ribbonlike flatworms dwelling in the gastrointestinal tract

Tubing—the act of passing a tube through a goat's esophagus to deliver milk or liquid medicine directly into the digestive tract

Udder—the female mammary system

Unrecorded grade—a goat, often of unknown ancestry, whose pedigree has not been recorded

Urinary calculi—stones formed in the urinary tract

Vaginal prolapse—protrusion of part or all of the vagina in late-gestation does

Wether—(noun) a castrated male goat of any age; (verb) the act of castrating a male goat

Whey—liquid left after removing curds from curdled milk

Wisconsin mastitis test (WMT)—an easy, do-it-yourself home mastitis test

Yearling—a goat of either sex between one and two years of age

Resources

Online Resources

The Internet is a goat keeper's best friend. Whatever information you seek, if you know where to look, it's out there. Here's a guide to some of the best goat Web sites in the world.

Breeders Directories

Most breed-specific sites offer directories, as do many of the all-purpose goat sites listed under Other Useful Web Sites.

Breeders' World Goat Directory

http://www.breedersworld.com
Breeders' World bills itself "The First Complete Online Livestock Breeders' Directory." Click on the *Goat Directory* link to access the directory, chat rooms, and a breeders forum; equipment and book suppliers pages; and a comprehensive list of goat registries and goat associations.

DMOZ Open Directory Project

http://dmoz.org/Business/Agriculture_and_Forestry/Livestock/Goats
The Open Directory Project is the largest human-edited directory on the World Wide Web. More than one thousand sites are cataloged in their goat resources directory. Visit to locate associations, breeders, supplies and equipment, recipes, and educational sites galore.

Dairy Goat Organizations

Alpines International Breed Club

http://www.alpinesinternationalclub.com
Established in 1958, the Alpines International Breed Club seeks "to

develop, preserve, and promote the French and American Alpine Dairy Goat." The organization sponsors Alpine specialty shows and offers production awards for Alpine dairy goats.

American Dairy Goat Association (ADGA)

http://www.adga.org

As of early 2005, the ADGA had registered 1,708,378 goats of eight breeds (Alpine, LaMancha, Nigerian Dwarf, Nubian, Oberhasli, Saanen, Sable, and Toggenburg); sanctioned 56,183 shows; and assigned 20,419 herd names. Access a comprehensive breeders directory and scores of helpful resources at the excellent ADGA Web site.

American Goat Society (AGS)

http://www.americangoatsociety.com

The AGS' purebred dairy goat registry was incorporated in 1936. It registers Alpines, LaManchas, Nigerian Dwarfs (it's this breed's primary registry), Nubians, Oberhaslis, Pygmies, Saanens, Sables, and Toggenburgs. Several excellent brochures, including "Beginners' Guide to Dairy Goats" and "ABC's of Milk Testing" are downloadable as PDF files via the AGS Web site. Click on *Affiliate Organizations* to find an AGS-affiliated dairy goat club in or near your locale.

American Nigerian Dwarf Dairy Goat Association (ANDDA)

http://www.andda.org

If you'd like to keep cute, small Nigerian Dwarf goats for dairy purposes rather than pets, this is your site! Click on Milk Production to see how incredibly milky these little goats can be when selected for milk production.

Canadian Goat Society (CGS)

http://www.goats.ca

The CGS registers the same breeds as its American counterpart does and administers many of the same type of programs.

Guernsey Goat Breeders of America (GGBA)

http://guernseygoats.org

Long-haired Guernsey dairy goats hail from the Isle of Guernsey off of England. They are a rare breed here and in their British homeland.

International Dairy Goat Registry (IDGR)

http://idgr.info/index

The IDGR registers dairy and rare goat breeds.

International Nubian Breeders Association (INBA)

http://www.i-n-b-a.org

Visit the INBA Web site to read about breed history, to find Nubian breeders in your locale, or to buy a nifty INBA T-shirt!

International Sable Breeder's Association (ISBA)

http://internationalsablebreeders association.vpweb.com

Visit the ISBA Web site to view Sable-related articles, photo albums, and show results as well as dozens of archived issues of the *Sable Journal* (excellent newsletters!), a directory of members and breeders, and a Sable pedigree database.

Kinder Goat Breeders Association (KGBA)

http://members.aol.com/kgbassn/index.htm
Articles about Kinder goat history and standards, a breeders directory, and a slew of other goat resources await you at the KGBA Web site.

Miniature Dairy Goat Association (MDGA)

http://miniaturedairygoats.com
Visit the MDGA Web site to learn about MiniLaManchas (Mini-Manchas), MiniAlpines, MiniNubians, Mini-Oberhaslis, MiniSaanens, and MiniToggenberg dairy goats.

National Miniature Goat Association (NMGA)

http://www.nmga.net
The NMGA Miniatures are small, triple-purpose goats suitable for dairying, meat production and pets.

National MiniNubian Breeders Club (NMBC)

http://www.mininubians.com
Find a breeder, research a pedigree, or buy MiniNubian merchandise at the NMBC Web site.

National Saanen Breeders Association (NSBA)

http://nationalsaanenbreeders.com
The NSBA Web site boasts a huge breeders directory, informative articles, and the latest news about these Swiss goats and the people who love them.

National Toggenburg Club (NTC)

http://nationaltoggclub.org
Visit the NTC Web site to view specialty show information as well as the latest list of Toggenburg Bell Ringer award winners.

Nigerian Dwarf Goat Association (NDGA)

http://www.ndga.org
The NDGA is one of several organizations devoted to registering and promoting the increasingly popular Nigerian Dwarf goat.

Ontario Goat Milk Producers' Association (OGMPA)

http://www.ontariogoatmilk.org
OGMPA's Web site is a treasure trove of resources for dairy goat aficionados, especially commercial producers. Don't miss it!

Dairy Goat Web Sites

Dairy Goat Manual

https://rirdc.infoservices.com.au/items/08-206
Visit and click on Download PDF 1.73 MB to download this $25, 78-page Australian book for free.

Fias Co Farm

http://fiascofarm.com/goats

This is a rich source of goat keeping information and recipes. Though slanted for dairy goat keepers, most of the material applies to goats of every kind.

GoatDairyLibrary.org

http://goatdairylibrary.org

Visit this amazing resource to access hundreds of pages of information. Some is dairy-oriented but much applies to goat keeping in general.

Saanendoah—Dairy Goat Information of the Serious Kind

http://www.saanendoah.com

No matter what type of goats you own, check out this site. Floppy kid syndrome, CAE, copper deficiency, caseous lymphadenitis, Johne's, colostrum supplementation, and do-it-yourself fecal testing are some of the serious topics explored. Additionally, take time to click on *The Liter Side of Goats* (it'll make you smile).

Fiber Goat Organizations

American Angora Goat Breeders Association (AAGBA)

http://www.aagba.org

Click on Photo Gallery to view some great historical images of Angora goats.

American Nigora Goat Breeders Association (ANGBA)

http://nigoragoats.homestead.com

Nigora goats are created by combining Nigerian Dwarf and Angora genetics. Visit this informative Web site to learn about these cute, small fiber goats. You can join this up-and-coming group and participate at their Yahoo group for free!

Colored Angora Goat Breeders Association (CAGBA)

http://www.cagba.org

The CAGBA Web site is a perfect first stop for anyone contemplating Angora goat ownership. Learn how and where to buy Angora goats, how to breed for color (the genetics resources are truly outstanding), and all about these beautiful goats for 4-H or FFA.

Mohair Council of America (MCA)

http://www.mohairusa.com

The MCA calls itself, "a one-of-a-kind organization exclusively dedicated to a one-of-a-kind commodity—mohair—the luxuriant fleece of the Angora Goat." Visit the MCA Web site to learn about Angora goats, fiber care, and the American mohair industry.

Pygora Breeders Association (PBA)

http://www.pygoragoats.org

Find a breeder or a 4-H fiber club in your area, subscribe to a Pygora e-mail list, or access the registry's online herdbook via this handsome and helpful PBA Web site.

Fiber Goat Web Sites

Angora Goats and Mohair Production

http://sanangelo.tamu.edu/publications/angora-goat-and-mohair-production

Visit Texas A&M's Angora goat site to download two free, colorful guides, "Angora Goats: A Shear Delight" and "Angora Goat and Moahair Production."

Capricorn Cashmere

http://www.capcas.com

Capricorn Cashmere hosts the best cashmere resource on the Web. Be sure to click on *4-H Center* and read "50 Ways to Kill a Goat"—it just might save your animals' lives.

Goat Fiber Information

http://textilelinks.com/spin/goatinfo.html

Follow these links to everything you want to know about mohair and cashmere.

Joy of Handspinning—Types of Mohair

http://www.joyofhandspinning.com/mohair.html

Visit this site for a great explanation of mohair and its uses and to view a short video featuring handsome Angora goats.

Your 4-H Angora Goat Project

http://4h.msue.msu.edu/uploads/resources/4H1480Your4-HAngoraGoatProject.pdf

Download Michigan State University's 36-page 4-H Angora goat guide. It's free!

Meat Goat Organizations

American Boer Goat Association (ABGA)

http://www.abga.org

Visit the ABGA Web site to find Boer breeders or a Boer goat show in your area, to locate goats or semen for sale, or to simply peruse a huge supply of interesting Boer goat material.

American Kiko Goat Association (AKGA)

http://www.kikogoats.com

Great Kiko goat articles, an AKGA breeders directory, membership list, registry news, and classified ads. Click on *Why Kikos?* to learn more about these brawny, efficient meat makers.

Canadian Meat Goat Association

http://www.canadianmeatgoat.com

Kids, click on CMGA Youth to download two great meat goat activity books—they're free!

Empire State Meat Goat Producers Association (ESMPGA)

http://www.esmgpa.org

Mouse over topics in the menu to access hundreds of resources, not only for meat producers, but also for goat fanciers in general. A best bet: find *Links*, then scroll to *Files to Download* and take your pick of more than seventy free goat-related PDF downloads. Don't miss "Basic Medications and Equipment for Dairy Goats"—it's a honey!

International Boer Goat Association, Inc. (IBGA)

http://www.intlboergoat.org
Check out the breeders directory, read classified ads, subscribe to the *Boer Breeder* bimonthly magazine, or buy handsome IBGA logo merchandise.

International Kiko Goat Association, Inc. (IKGA)

http://www.theikga.org
The IKGA registers and promotes Kiko goats. Download their free brochure—it's a good one! Prospective meat goat producers, no matter their favorite breed, will benefit from the IKGA's fifteen-page PDF bulletin, "Hints for the Inexperienced Goat Farmer." Click on *Publications* to download it.

National Kiko Registry (NKR)

http://www.nationalkikoregistry.com
Click on Kiko Information to access a dozen informative articles about meat goat production in general and Kiko goat in particular.

North American Savannah Association (NASA)

http://savannahassociation.com
If you're interested in Savannah goats NASA's web site is a treasure trove of information, including several not-to-be-missed educational videos.

Spanish Goat Association (SGA)

www.spanishgoats.org
Spanish goats are an endangered breed of exceptionally hardy meat goats.

United States Boer Goat Association (USBGA)

http://www.usbga.org
Read the latest news, peruse the classifieds, download past issues of the USBGA magazine in PDF format, or look up a USBGA pedigree online (use of the database is free).

Meat Goat Web Sites

Bar None Meat Goats

http://www.barnonemeatgoats.com
This informative site is the Web home of Robin Walters, contributor of many of the great Advice From the Farm tips in this book. Don't miss *Raising Market Goats* and *Goat Information Page*!

Boer & Meat Goat Information Center

http://www.boergoats.com
You'll find articles galore, USDA livestock reports, show results, shipping regulations, producers directories, even a photo contest.

Feasibility of Goat Production in West Virginia: A Handbook for Beginners

http://www.ca.uky.edu/anr/Agent%20Resources/pdf/WV%20goat%20pub.pdf
No matter where you live, you'll reap heaps of information from the pages of this 34-page handbook.

Jack & Anita Mauldin's Boar Goats

http://www.jackmauldin.com
Interested in raising Boer Goats? Click

on *Quick Start* to learn all the particulars, then *Industry Info* for the rest of the story. All goat owners will appreciate the articles under *Goat Health Information* and *Management Informa-tion*. The *Related Goat Sites* links page is superb!

MAC Goats

http://members.psyber.com/macgoats
Scroll down Matt and Claudia Marcus-Gurn's home page and click on *Education* to access a huge collection of educational goodies. Although some articles will interest mainly meat goat producers, most will appeal to owners of other types and breeds as well. Scope out the beautiful show buck, Downen R33 Hoss, while you're there—he's Salem and Shiloh's famous daddy!

Onion Creek Ranch Tennessee Meat Goats

http://www.tennesseemeatgoats.com
Suzanne W. Gasparotto's site is simply fantastic! Click on *Articles* to access oodles of valuable items. Don't miss these three: "Do Your Own Fecals," "Supplies Every Goat Rancher Needs," and "Goat Medications and How to Use Them."

Other Breed Organizations

GoatWorld Goat Registry

http://www.goatregistry.com
If your pet or packgoat isn't eligible for registration in the mainline herdbooks, but you'd like to have papers for identification purposes, the GoatWorld Goat Registry will register your caprine sidekick.

International Fainting Goat Association (IFGA)

http://www.faintinggoat.com
The IFGA registers all types of Myotonic (fainting) goats. Visit this site to learn about these interesting, ALBC-listed heritage goats.

Miniature Silky Fainting Goat Association (MSFGA)

www.msfgaregistry.com
The Miniature Silky Fainting Goat is a small, long-haired pet breed. These cute goats resemble Silky Terrier dogs.

Myotonic Goat Registry (MGR)

http://www.myotonicgoatregistry.net
Everything you wanted to know about Myotonic (fainting) goats—it's here, including great articles, a breeders' directory and a pedigree database.

National Pygmy Goat Association (NPGA)

http://www.npga-pygmy.com
Locate the many Pygmy goat shows, breeders, and affiliated clubs through the NPGA Web site. While you're there, check out the many outstanding goat articles that are accessible by clicking *Goat Resources*.

Pedigree International LC

http://www.pedigreeinternational.com
Pedigree International LC provides herdbooks and breeders directories for producers of rare and composite livestock breeds; among them Kiko,

GeneMaster, Myotonic, Tennessee Meat Goats, Savannah, Spanish, TexMaster, and Cashmere goats.

North American Heritage Livestock Breed Conservancies

American Livestock Breeds Conservancy (ALBC)

http://www.albc-usa.org
The ALBC promotes and helps preserve nearly one hundred breeds of heritage cattle, goats, horses, asses, sheep, swine, and poultry. Visit the site to see how you can help.

Heritage Breeds Conservancy (HBC)

http://www.nehbc.org
The HBC works to preserve historic and endangered livestock and poultry breeds. Conservators from the U.S. and Canada are listed in the HBC Directory. An online forum, classified ads, and nice selection of conservation-related links round out this informative Web site.

Rare Breeds Canada (RBC)

http://www.rarebreedscanada.ca
Like its American cousin, the American Livestock Breeds Conservancy, RBC preserves and promotes heritage livestock, including goats.

San Clemente Island Goat Association (SCIGA)

www.scigoats.org
San Clemente Island goats are a critically endangered breed. If you're looking for a conservation project, this is your goat.

Other Useful Web Sites

There are hundreds, perhaps thousands, of privately hosted goat information sites on the Internet. We apologize to the owners of other exceptional sites we've visited—we wish we had room for them all!

4-H Harness Goat Project

http://ansci.cornell.edu/4H/goats/harness_goat.pdf
You'll appreciate this 57-page harness goat guide even if you aren't a kid.

Amber Waves Pygmy Goats

http://amberwavespygmygoats.com/index.php
Use the pull-down menu to access a fantastic amount of information, as well as useful interactive tools such as dosage, due date and pedigree calculators.

Biology of the Goat

http://www.goatbiology.com
Karin Christensen's animations make understanding goat biology a relative breeze. Visit to view sample animations and download her animated goat gestation calculator. Her *Biology of the Goat* CD is fantastic!

British Feral Goat Research Group

http://britishferalgoat.org.uk
Interested in caprine history and preservation efforts? Scope out this great site.

British Goat Society (BGS)

http://www.allgoats.com
Founded in 1879, the BGS brings you scores of great goat-keeping features.

Cyber Goats

http://www.cybergoat.com
Classified ads and auctions, links to breeders and resources, breed info and breed-specific e-groups, and a helpful guide to caprine veterinarians.

Dreamgoat Annie

http://www.dreamgoatannie.com
Visit my web site to download lots of goat material in PDF format.

Goat Connection—The Whole Goat Catalog Online

http://goatconnection.com
Use the pull-down menu at the *Goat FAQ Database* to read scores of great articles archived at Goat Connection, peruse its "Name That Goat" feature to name your new kids, mail a goat-themed e-card, or shop for goat-goody gifts.

Goat Kingdom—Home of Duh Goat Man

http://duhgoatman.tripod.com
Visit Duh Goat Man for a mind-boggling collection of links to breeders, information resources, supplies, periodicals, and everything else to do with goats.

Goat Wisdom

http://www.goatwisdom.com
Goat Wisdom is a virtual one-stop information center. Check it out.

GoatWorld

http://www.goatworld.com
Visit GoatWorld to access the most extensive collection of goat articles on the Internet. It hosts Goat911, where you can contact goat experts 24-7 to request free assistance with your sick or injured goat. Click the *Emergency 911* logo and follow the prompts.

High Uinta Pack Goats

http://www.highuintapackgoats.com
High Uinta Pack Goats brings you pack goat selection, training, and maintenance articles, tips, and FAQs. Don't miss this site—it's a good one!

International Goat Association (IGA)

http://www.iga-goatworld.org
IGA is an international organization of scientists, educators, goat producers, vets, and others who promote sustainable goat management and the sale of goat products to improve human nutrition and social welfare.

Kinne's Minis

http://kinne.net
Kinne's Minis are Pygmy goats. Like Amber Waves, this is a first-class place to access a number of goat articles.

Maryland Small Ruminant Page

http://www.sheepandgoat.com/organ.html
Visit the Maryland Small Ruminant Page in order to locate additional

organizations including scores of state and regional dairy, meat, and fiber goat associations.

North American Pack Goat Association (NAPgA)

http://www.napga.org
The NAPgA promotes wilderness packing with packgoats. Click on *NAPgA Informational Items* to download a packgoat brochure in PDF format and to join the NAPgA e-mail list via the Web site.

YahooGroups Goat-Oriented E-Mail Lists

http://groups.yahoo.com
The easiest to use and access goat listservs on the Internet are those hosted by YahooGroups, which hosts at least 800 goat-related e-mail groups. Find them by typing *goats* in the search box,. Or narrow the search by using multiple-word searches such as *Saanen goats, goats Minnesota, Livestock Guardian Dogs*, or *meat goats*. Some YahooGroups we've sampled and enjoyed are GoatBiology, practical-goats, goats_101, packgoat, AlpineTalk, the_boer_goat, chevontalk, homedairygoats, and MissouriMeatGoats.

Herd Guardians

Dog Owner's Guide: Livestock Guard Dogs

http://www.canismajor.com/dog/livestck.html
Visit to learn about livestock guardian dog breeds, including breed-specific behavior issues.

Livestock Guard Dogs, Llamas and Donkeys

http://www.nal.usda.gov/awic/pubs/1218.pdf
Not sure which species is best for you? This government might help you decide.

Livestock Guardian Dogs (LGDs)

http://www.lgd.org
Here you'll find everything you need to know about livestock guardian dogs. Subscribe to the LGD-Lovers (LGD-L) e-mail list; browse the list's LGD FAQs; and click on *Library* to peruse hundreds of informative articles. This site is brought to you by the Livestock Guardian Dog Association—don't miss it, it's great!

LlamaOrg

http://www.llama.org
Put a llama in your life! Surf this comprehensive site to learn everything you need to know before buying llamas. Scroll down the home page to "What Do You Do With a Llama" and click on *Guarding Livestock* for the skinny on llamas as herd guardians.

Using Guard Animals to Protect Livestock

http://icwdm.org/PDF's/MOguardanimals1996.pdf
This 14-page download is a best-bet source of essential information on choosing and using a guardian dog, llama or donkey.

Recipes

Caprine Cookbook at Goat Connection

http://www.goatconnection.com/recipes

For the greatest selection of goat recipes on the Web, visit the fabulous "Caprine Cookbook at Goat Connection" for hundreds of recipes incorporating chevon and cabrito and various goat's milk products.

Goat Cuisine – A Culinary Delight

http://www.aces.edu/pubs/docs/U/UNP-0123/UNP-0123.pdf

Alabama Cooperative Extension brings you this 8-page guide to cooking with goat meat including recipes for Goat Burgers, Creole Cabrit, Slow Goat in a Curry and more.

Goat Meat Recipes at Copeland Valley Farms

http://www.goatmeats.com/asp/Information/GoatMeatRecipes.asp

The goat meat producers at Copeland Valley Farms bring you mouth-watering recipes including Chevon Stew, Southwest Leg of Goat and Chevon in a Pit!

Goat Milk Recipes from the MEYENBERG Kitchen

http://www.meyenberg.com/cooks

Mmm-mmm, this page features twenty tasty recipes for goat's milk ice creams, yogurt, whipped cream, fudges, custards, and casseroles. Click on *Resources* to download the company's free "Goat Milk Gourmet Cookbook."

Goat Recipes at Fiery Foods

http://www.fiery-foods.com/recipesearch/category/goat

If you crave fiery food, you'll love Stoba di Cabrito (Curaçao-Style Kid Stew), Trinidadian Coconut-Curried Goat, Southwestern Cabrito, and several more tongue-sizzling recipes at this page.

Jack & Anita Mauldin's Goat Recipes

http://www.jackmauldin.com/goat_recipes.htm

The Mauldin's outstanding meat goat information site brings you twenty-eight yummy goat recipes, from Texas Ranch Style Gumbo to Jamaican Curried Goat, along with suggestions for goat stews, sausage, stir fries, and cabrito loaf.

Recipes at Fias Co Farm

http://fiascofarm.com/recipes

Visit the nice folks at Fias Co Farm and learn to make eight cheeses including feta, mozzarella, chevre, and queso blanco.

Supplies

American Livestock Supply (ALS)

http://www.americanlivestock.com

800-356-0700

ALS stocks a full line of goat (and sheep, cattle, poultry, swine, etc.) vaccines and equipment at discount prices. Order online or request a free print catalog.

Caprine Supply

http://www.caprinesupply.com
800-646-7736
Caprine Supply stocks everything goat. Click on *Helpful Information* to read two dozen excellent articles about goat management and dairying, and call or fill out an online form to request Caprine Supply's great free catalog.

Hoegger Goat Supply

http://www.hoeggerfarmyard.com
800-221-4628
Hoeggers' carries a fantastic array of goat supplies. Click on *The Farmyard* to access a slew of helpful articles, and call or fill out the online catalog request form to receive your copy of Hoeggers' free eighty-page catalog.

Jeffers Livestock Supply

http://www.jefferslivestock.com
800-533-3377
Jeffers offers the same wide selection of livestock equipment and pharmaceuticals as American Livestock Supply, at competitive prices. Jeffers' catalogs are free.

Mid-States Wool Growers Cooperative Association

http://www.midstateswoolgrowers.com
East of the Mississippi: 800-841-9665
West of the Mississippi: 800-835-9665
Click on *Online Store,* then on *Lambing & Kidding Supplies* or *Sheep & Goat Equipment* to access the Mid-States Wool Growers Cooperative Association's outstanding selection of goat essentials, including hard-to-find items such as CL-Nanny Replacer Colostrum and adorable Kid Jammies.

New England Cheesemaking Supply Company

http://www.cheesemaking.com
New England Cheesemaking Supply Company markets books, failure-resistant cheese-making kits, and everything else you need to make tasty goat cheese.

Northwest Pack Goats & Supplies

http://www.northwestpackgoats.com
888-PACKGOAT
A 28-page catalog features a nice selection of goat supplies including halters, goat coats, first aid kits, goat-themed rubber stamps, books, and pack goat gear. Don't miss how-to training articles at the Web site as well as those scattered throughout the free print catalog.

Premier 1 Supplies

http://www.premier1supplies.com
800-282-6631
Premier has provided shepherds and goat keepers with fencing, clippers and shearers, ear tags, and expert advice for more than twenty-five years. Shop its comprehensive online store or get the free catalog that's packed with sheep and goat products, fencing, and clipper and shearing machines.

Quality Llama Products, Inc.

http://www.llamaproducts.com
800-638-4689
In addition to llama and alpaca supplies, Quality Llama Products, Inc., carries a treasure trove of equipment and books for fanciers of miniature donkeys, miniature horses, reindeer, and goats. You'll find information on topics such as halters, harnesses, carts, goat bits, and goat packing gear. Their catalog is free.

Sullivan Supply

http://www.sullivansupply.com
Texas warehouse: 800-588-7096
Iowa warehouse: 800-475-5902
If you show your goats (or simply spiff them up for special occasions), you need the free Sullivan catalog that's full of grooming tools and gadgets, blankets, stands, shampoos, and conditioners.

Valley Vet Supply

http://www.valleyvet.com
800-419-9524
The free Valley Vet Supply catalog features a huge selection of livestock supplies, equipment, a d pharmaceuticals at attractively discounted prices.

University Resources

Most state universities and state extension services distribute papers and bulletins of interest to goat keepers. While visiting these Web sites, check for useful bulletins under headings such as *Farm Construction, Forage* (hay and pasture), and *Poisonous Plants*. A lot of the information in sheep bulletins also applies to goat keeping. To compile an up-to-date library of goat materials, download appropriate PDF files to save for future reference, print favorite bulletins and file them, or bind printouts to create your own goat reference handbook.

Agricultural Marketing Resource Center (AgMRC)

http://www.agmrc.org
AgMRC is composed of marketing experts from Iowa State University, Kansas State University, and the University of California who work together to create and disseminate information about value-added agriculture. Don't miss this valuable site! Explore marketing trends and peruse thousands of valuable print and online resources. Access goat resources by clicking on *Commodities & Products*, then on *Livestock*, then on *Goats*.

Alabama Cooperative Extension System

http://www.aces.edu
The Alabama Cooperative Extension System is an especially rich source of goat management resources. To access them, click on *Agriculture*, then *Sheep & Goats*.

Clemson University Extension (South Carolina)

http://www.clemson.edu/extension
To access goat materials, click on *Resources*, then *Extension*, then *Farm*, and finally, *Goats*.

Cornell University 4-H Youth Programs

http://www.ansci.cornell.edu/4H/goats/index.html
Even if you're not a kid, check out Cornell University's 4-H goat project pages. Click on Dairy Goats or Meat Goats to access loads of useful material and check out the castrating, ear tagging, disbudding and tattooing skill slide shows while you're there.

Cornell University Goat Management

http://ansci.cornell.edu/goats
This fantastic site is a must-visit for all goat keepers. Click on Resources, Links and Market Info for more useable information than you can possibly imagine.

Cornell University's Sheep & Goat Marketing Directory

http://sheepgoatmarketing.info/education.php
Click on Education to access an astounding amount of information about marketing chevon and lamb.

Langston University Research and Extension (Oklahoma)

http://www.luresext.edu
Click on the goat logo to access Langston University's E (Kika) de la Garza Institute for Goat Research, one of the best goat resources on the World Wide Web.

Maryland Small Ruminant Page

http://www.sheepandgoat.com
Susan Schoenian, Sheep and Goat Specialist for the University of Maryland Cooperative Extension, hosts this collection of original documents and links to thousands of online resources.

Mississippi State University Extension Service

http://msucares.com
Click on *Livestock* then *Sheep & Goats.* Look for *Publications* in the right-hand menu. Click on it, and you're there!

Oklahoma State University's Breeders of Goats

http://www.ansi.okstate.edu/breeds/goats
This site features information and pictures of hundreds of breeds worldwide.

Oklahoma State Univeristy's Meat Goat Manual

http://meatgoat.okstate.edu/oklahoma-basic-meat-goat-manual-1
This is an excellent resource for all goat keepers; each chapter is downloadable.

Pennsylvania State University Meat Goat Home Study Course

http://extension.psu.edu/courses/meat-goat
The study guides for this course are available online for free!

Purdue University's Dairy Goats and Goat Links

http://www.ansc.purdue.edu/goat/goatlink.htm
Visit Purdue University's dairy goat web site to view links, articles and webinar videos online.

Purdue University's Meat Goats @ SIPAC

http://www.ansc.purdue.edu/caprine
This Purdue University site is a treasure trove of information for meat goat producers. Click on Extension Publications to access useful articles and Links to visit similar sites.

Texas Cooperative Extension Bookstore

http://tcebookstore.org
Meat goat producers of all ages will like Texas A&M's comprehensive 4-H meat goat handbook. To download it, enter *Goats* in the search box, then click on *4-H Meat Goat Guide.*

University of Kentucky Goat Research and Education

http://www.uky.edu/Ag/AnimalSciences/goats/goat.html
The University of Kentucky goat site is jam-packed with useful information including publications, copies of newsletters, videos and oodles of links.

University of Maryland Sheep & Goat Program

http://extension.umd.edu/sheep-goats
Check out the program's Wild & Wooly newsletter archives and the interesting section about worm control

University of Maryland's Western Maryland Research and Education Center

http://www.westernmaryland.umd.edu
Here, accessing a huge selection of goat resources is simple: click on *Sheep & Goats*, and there you are! Peruse or download *Maryland Sheep and Goat Producer* newsletters, and access the Maryland Small Ruminant Page, the Northeast Sheep and Goat Marketing Web site, and a lot of other useful goat-oriented materials via this Web site.

University of Missouri Extension

http://muextension.missouri.edu
Although University of Missouri Extension doesn't offer goat-specific publications, its agricultural engineering (fencing, farm structures, livestock equipment) bulletins are outstanding. Many of the school's excellent sheep bulletins will interest goat owners, too. To access goat resources, click on *Agriculture*, then *Animals*, and then *Goats.*

Washington State University Small Farms Teams – Goats

http://smallfarms.wsu.edu/animals/goats.html
Click on the types of goats that interest you most.

Veterinary Associations

American Association of Small Ruminant Practitioners (AASRP)

http://www.aasrp.org
The AASRP is a professional organization for veterinarians and veterinary students interested in small ruminant medicine.

American Holistic Veterinary Medical Association (AHVMA)

http://www.ahvma.org
Browse the AHVMA referral directory to find licensed holistic veterinarians in your area. A click on *Links to Veterinary Resources* leads you to specialty veterinary medical organizations such as the Academy of Veterinary Homeopathy, the American Veterinary Chiropractic Association, and the Veterinary Botanical Medicine Association.

American Veterinary Medical Association (AVMA)

http://www.avma.org
Visit the AVMA to view or download articles written by vets for the layman.

Government Resources

United States

Agricultural Marketing Service @ USDA (AMS) and the National Organic Program

http://www.ams.usda.gov
If you'd like to market organic chevon or cabrito and are seeking the official word on organic certification, visit the AMS Web site. Click on *National Organic Program*, then your topic of interest.

USDA Animal and Plant Inspection Services Scrapie Program

http://www.aphis.usda.gov/animal_health/animal_diseases/scrapie
In most cases, owners of goats kept on the same premises as sheep must participate in the USDA's mandatory Scrapie-Eradication Program. Since rules seem to change at lightning speed, check this site for the latest skinny. If you're still confused, peruse a frequently updated, reader-friendly explanation, "Goat Scrapie Program Information," at http://www.goatworld.com/articles/goatscrapie.shtml

Canada

British Columbia Ministry of Agriculture and Lands

http://www.agf.gov.bc.ca
The Government of British Columbia's huge collection of excellent goat bulletins can be accessed by clicking on *Reports & Publications*, then *Goats*, and then *Fact Sheets & Publications*.

Ontario Ministry of Agriculture, Food, and Rural Affairs

http://www.omafra.gov.on.ca/english
This site is a rich source of information for goat owners and producers. To access hundreds of management titles,

click on *Agriculture*, then on *Livestock*, then scroll down the page to *Goats*.

Saskatchewan Agriculture and Food

http://www.agr.gov.sk.ca
Find excellent archived goat documents by clicking on *Livestock*, then on *Sheep and Goats*.

Australia

New South Wales Department of Primary Industries

http://www.dpi.nsw.gov.au
To peruse the excellent goat resources, click *Agriculture NSW*, then *Livestock*, and then *Goats*.

Periodicals

General Interest

United Caprine News

http://www.unitedcaprinenews.com
Printed in newspaper format, *United Caprine News* is a leading source of up-to-date information on the care, management, and feeding of goats. Each monthly issue is loaded with information on all aspects of goat keeping, including health, feeding, breeding, housing, showing, milking, and routine care.

Dairy Goat Periodicals

Dairy Goat Journal

http://www.dairygoatjournal.com
Each issue of *Dairy Goat Journal* features timely articles about raising, breeding, and marketing dairy goats; health issues; and recent news of interest to goat owners and the dairy goat industry. Click on *past issues* to view a plethora of archived articles.

Ruminations

http://www.smallfarmgoat.com
Ruminations is a well-written, bimonthly magazine for Nigerian Dwarf and other Miniature Dairy Goat enthusiasts. A compendium of articles from back issues is available through the Ruminations Web site.

Meat Goat Periodicals

Goat Rancher

http://www.goatrancher.com
An attractive monthly, *Goat Rancher* reports the latest news in the production, health and management, and marketing of meat goats by featuring articles and timely tips from established producers of Boer, Kiko, Spanish, and Tennessee Meat Goats. At the Goat Rancher Web site, click on *Markets* to check up-to-date prices at dozens of goat auctions throughout the United States.

Meat Goat Monthly News

http://www.ranchmagazine.com
Meat Goat Monthly News is a publication about the U.S. meat goat industry. For subscription information, visit the Ranch & Rural Living magazine web site and click on *Meat Goat Monthly News*.

Recreational Goat Periodicals

Goat Tracks

http://www.goattracksmagazine.com
Goat Tracks is a well-written quarterly

for owners and admirers of pack and harness goats.

Books

There are many great goat books on the market—more than we can list in this brief bibliography. These are our in-print favorites, readily available via Hoeggers, Caprine Supply, Quality Llama Products, or Amazon (http://www.amazon.com). But we love older goat books, too; so, for a truly well-rounded library, we suggest you look to eBay (http://www.eBay.com) for out-of-print; antiquarian; and British, Australian, or foreign-language goat books

General Interest

Beberness, Alice and Eddy, Carolyn. *Field First Aid for Goats*. Eagle Creek Packgoats, 2008.

Coleby, Pat. *Natural Goat Care.* Acres USA, 2001.
Pat Coleby explains holistic goat keeping from every conceivable angle, in 371 fact-filled pages.

Grant, Jennie Palches. *City Goat: The Goat Justice League's Guide to Backyard Goat Keeping*. Mountaineers Books, 2012.

Jaudas, Ulrich. *The New Goat Handbook.* Barron's, 1989.
The author packs an amazing amount of information into this slim, 93–page volume. The book's glorious illustrations and color photographs are superb.

Kindsedt, Paul. *American Farmstead Cheese: The Complete Guide to Making and Selling Artisan Cheeses.* Chelsea Green, 2005.
If you're an experienced cheese crafter who dreams of marketing artisan cheeses, don't miss this comprehensive, 276-page hardcover cheese-making guide. New England Cheesemaking Supply calls it, "The best book we have seen to date."

Mionczynski, John. *The Pack Goat.* Reavis, 2004.

Mowlem, Alan. *Goat Farming.* 2d ed. Farming Press, 1992.
This handy 208-page hardcover guide covers profitable dairy, meat, and fiber production from a British perspective.

Sayer, Maggie. *Storey's Guide to Raising Meat Goats.* 2d ed. Storey Publishing, 2010.

Sinn, Rosalee. *Raising Goats for Milk and Meat: A Heifer Project International Training Course.* Heifer Project Intl., 1992.
This 110-page, spiral-bound book covers the basics of goat keeping. Included: 36 pages of record-keeping sheets.

Smith, Cheryl K. *Raising Goats for Dummies.* For Dummies, 2010.

Weaver, Sue. *The Backyard Goat: An Introductory Guide to Keeping Productive Pet Goats.* Storey Publishing, 2011.

Weaver, Sue. *Storey's Guide to Raising Miniature Livestock: Goats, Sheep, Donkeys, Pigs, Horses, Cattle, Llamas.* Storey Publishing, 2011.

Dairy Goat and Dairy Product Books

Carroll, Ricki. *Home Cheese Making: Recipes for 75 Delicious Cheeses.* 3d ed. Storey, 2002.
Authored by Ricki Carroll of the New England Cheesemaker's Supply, this handsome, easy-to-use, 224-page cheesemaking manual features eighty-five home-crafted cheese and dairy product recipes. John and I made yummy cheese using its clear instructions—if we can, so can you!

Le Jaouen, Jean-Claude. *The Fabrication of Farmstead Goat Cheese.* Cheesemaker's Journal, 1990.
Written for experienced cheese makers, *The Fabrication of Farmstead Goat Cheese* is the ultimate guide to crafting artisan-quality goat milk cheeses.

Luttmann, Gail. *Raising Milk Goats Successfully.* Williamson, 1986.
This 176-page softcover manual has stayed in print so long because it's such an outstanding guide to home dairying, beloved by beginners and experienced goat keepers alike. It's my favorite dairy goat guide.

Mont-Laurier Benedictine Nuns. *Goat Cheese: Small Scale Production.* New England Cheesemaking Supply, 1983.
This eighty-eight–page paperback includes directions for making starter culture; presents several methods for testing acidity; and includes recipes for crafting goat cheese, butter, and yogurt.

Stewart, Patricia Garland. P*ersonal Milkers; a Primer to Nigerian Dwarf Goats.* Garland-Stewart Publishing, 2008.

Fiber Goat Books

Drummond, Susan Black. *Angora Goats the Northern Way.* 3rd ed. Stony Lonesome Farm, 1991.
You'll find everything you need to know to keep Angora goats, to shear them, and to market mohair in this comprehensive, 239-page softcover book. Both the Mitchams' book (below) and this one are excellent values, but this is my favorite of the two.

Mitcham, Stephanie, and Allison Mitcham. *The Angora Goat: Its History, Management and Diseases.* 2d ed. Crane Creek, 1999.
Every Angora goat keeper needs this book. Besides covering every aspect of Angora goat management, the Mitchams discuss livestock guardian dogs, herding dogs, and a lot of other topics other authors usually miss.

Meat Goat Books

Bowman, Gail B. *Raising Meat Goats for Profit.* Bowman Communications, 1999.
This 256-page treasure is an easy-to-read

introduction to every aspect of profitable meat goat production. Prospective chevon and cabrito producers: you need it!

Mitcham, Stephanie, and Allison Mitcham. *Meat Goats: Their History, Management and Diseases.* Crane Creek, 2000.
Written by the authors of *The Angora Goat: Its History, Management and Diseases*, this comprehensive manual takes a novel approach to the meat goat industry by suggesting Angora crosses, not Boer crosses make the best-tasting chevon.

Tomlinson, Sylvia. *The Meat Goats of Caston Creek.* Redbud, 1999.
This charming 181-page book is a laid-back anthology of stories, essays, and tips on raising meat goats. Included: an interesting, down-home recipe section titled "Cocinar Chevito" (cooking young goat).

Miscellaneous Books

Boldrick, Lorrie, and Lydia Hale. *Pygmy Goats: Management and Veterinary Care.* All Pub, 1996.
If you keep Pygmy goats, you absolutely need this 237-page softcover manual. You might also want to buy a copy for your Pygmies' vet, too. Owners of other breeds will also learn a lot from this great book. It's one of our favorites—don't miss it!

Dawydiak, Orysia, and David E. Sims. *Livestock Protection Dogs: Selection, Care, and Training.* Alpine Blue Ribbon Books, 2004.
This paperback is currently the only professionally published all-breed LGD book in print. From breed profiles (including breeds you might not know of) to material on selecting a puppy or older dog and training it as a guard or family companion—this 224-page book delivers.

Dunn, Peter. *The Goatkeeper's Veterinary Book.* 3rd ed. Farming Press, 1998.
This 210-page British book is likely the only caprine veterinary manual you'll ever need. It's especially strong on preventative medicine, making it a must-have for every goat keeper's bookshelf.

Eddy, Carolyn. *Diet for Wethers: Guide to Feeding Your Wether for Health and Longevity.* Eagle Creek Packgoats, 2001.
If you keep wethers for driving or packing or as pets, you'll definitely want a copy of this fact-filled 102-page book. It's the first (and only that we're aware of) how-to handbook written specifically for the wether goat owner.

—. *Practical Goatpacking.* Eagle Creek Packgoats, 1999.
New and experienced goat packers alike will benefit from the wealth of how-to material in this 142-page, softcover book. The Northwest Pack Goats supply catalog calls it "a thorough, nuts-and-bolts guide to the art and science of goatpacking. It will answer most of the questions

that arise for the beginning goatpacker and goat keeper." We agree!

Ekarius, Carol. *How to Build Animal Housing: 60 Plans for Coops, Hutches, Barns, Sheds, Pens, Nestboxes, Feeders, Stanchions, and Much More.* Storey, 2004.
This isn't a goat book per se, but if you keep goats, you need it. In 272 packed pages, Ekarius explains everything you need to know to plan and build livestock housing and equipment. Beginners and experienced builders alike will benefit from this book.

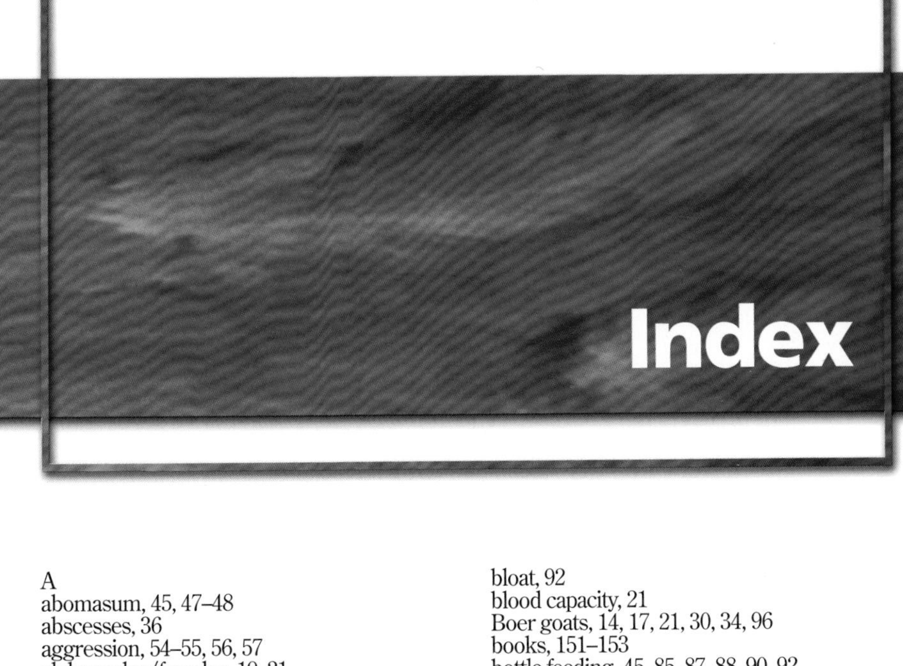

Index

About the Author

Sue Weaver is author of *Sheep: Small-Scale Sheep Keeping for Pleasure and Profit* and *Chickens: Tending a Small-Scale Flock for Pleasure and Profit*. She also has written hundreds of articles about animals over the years and is a contributing editor of *Hobby Farms* magazine. Sue and her husband, John (who provided most of the photos for this and her other books), live near Mammoth Spring, Arkansas, where as the proprietors of Ozark Thunder Boers they raise show-quality full-blood Boer goats. They also raise double-registered miniature American Brecknock Hill Cheviot and Keyrrey-Shee sheep, AMHR Miniature Horses of cob type, and American Curly horses.